D1015888

Power of Speech

Power of Speech

A History of Standard Telephones and Cables 1883–1983

Peter Young

London
GEORGE ALLEN & UNWIN
Boston Sydney

**George Allen & Unwin (Publishers) Ltd,
40 Museum Street, London WC1A 1LU, UK**

George Allen & Unwin (Publishers) Ltd,
Park Lane, Hemel Hempstead, Herts HP2 4TE, UK

Allen & Unwin, Inc.,
9 Winchester Terrace, Winchester, Mass. 01890, USA

George Allen & Unwin Australia Pty Ltd,
8 Napier Street, North Sydney, NSW 2060, Australia

First published in 1983

British Library Cataloguing in Publication Data

Young, Peter
Power of speech: a history of Standard Telephones and Cables 1883–1983.
1. Standard Telephones and Cables plc – History
I. Title
338.4'762138'0941 HD9696.T444.S/
ISBN 0–04–382039–5

Library of Congress Cataloging in Publication Data

Young, Peter.
Power of speech.
1. Standard Telephones and Cables (London, England) – History. 2. Telecommunication – Great Britain – History. 3. Telephone – Great Britain – History. I. Title.
HE8100.S725Y68 1983 384.6'065'41 82–11668
ISBN 0–04–382039–5

Set in 11 on 12pt Palatino by
Bedford Typesetters Ltd
and printed in Great Britain by
Mackays of Chatham

Contents

List of Illustrations

16 & 17 At Oxford, in 1957, a smaller crew did the same job; while paying out optical fibre cable between Hitchin and Stevenage in 1976, manoeuvring the whole reel needed only two men.

18 At Southampton docks, submarine cable made in STC's adjacent plant is loaded into the hold of a cable-laying vessel.

19 Surgically clean conditions are necessary in 'the dairy' at Greenwich, where repeaters are made to operate undisturbed for twenty-five years on the seabed.

20 The world's first undersea optical cable, manufactured by STC, being floated ashore in 1980 for its trial at Loch Fyne, a location chosen for its relatively deep, tidal salt water.

21 A TXE4A exchange is installed using STC's unique MITER system. MITER (Modular Installation of Telecommunications Equipment Racks) enables equipment to be assembled and tested at the factory. Once inside the exchange building, air cushions are used to move the units to their final position.

22 System X equipment in London's Baynard House exchange. System X – the fully digital switching system on which the modernisation of the UK network is being based – was developed jointly by STC, GEC, Plessey and British Telecom. STC withdrew from the project in October 1982.

23 Optical fibres as fine as human hairs are produced at STC's optical cable manufacturing unit at Harlow. One end of a glass preform is introduced into the pulling tower, heated to melting point and drawn under steady tension onto drums at the lower end.

24 J. E. Kingsbury, first agent and managing director of the Western Electric Company in London. Though he relinquished an active role in 1909, shortly after this photograph was taken, he remained a director until the change of ownership in 1925. Up to 1915 he used his spare time to write the book for which he is best known, the authoritative *The Telephone and Telephone Exchanges*.

25 Sir Frank Gill was a telephone pioneer who started 'down manholes and up poles'. He became not only chairman of STC, but also an internationally respected figure in telecommunications.

26 Sir Thomas Spencer began life as a Woolwich lad with a passion for engineering – and football. He lied about his age to join Western Electric at 14 in 1907. By 1936 he was sole managing director, and embodied the spirit of 'The Standard' through the war and into the 1950s.

27 Sir Kenneth Corfield, who joined STC as managing director in 1970 and has been chairman and chief executive since 1979.

1

The Bell Epoch

In *The Way We Live Now*, published in 1875, Trollope critically portrayed a Victorian society past its heyday, a society that manoeuvred money but that had lost its sense of direction. His characters were fictional but the mood he expressed was real. Economic historians were to look back on the years between 1873 and 1896 as the Great Depression, a period when Britain's economy stagnated into relative decline. Competition was coming from new nation states, in particular the USA and Germany, which had settled their physical frontiers and were turning their energies to technical and economic development.

Britain saw the solution to its trade problems in the building of an Empire on which the sun would never set. Trade would follow the flag. Capital was exported to develop sources of cheap food and raw materials and at the same time create captive markets for whatever the home country wanted to produce. While imperial ambition was growing in Britain others, especially in the USA, were concerned with exploiting scientific discoveries and creating markets for new products. Their progress was through a readiness to capitalise on new ideas. Enos M. Barton, a telegraph operator who rose to become a founder and president of one of America's biggest manufacturing companies, Western Electric, looked back:

> Fortunes had been made in developing the telegraph system, and it was the general expectation that there were other fortunes awaiting the development of new inventions. The electrical inventor could easily get the ear of the capitalist, and the capitalist even sought out the inventor. The decade from 1870 to 1880 saw the beginning of many great things in electricity. . . . The decade came in with short line printing telegraphs under development. It went out with the telephone, the electric light and the dynamo machine in full commercial swing.

Much of the scientific foundation for these developments was British, the work of Faraday, Cooke, Wheatstone, Clerk Maxwell and Swan. From a national point of view the invention of the telephone was an early example of the brain drain, though in an unscientific society few saw it as the basis for industrial growth. After losing two brothers in England through tuberculosis, Alexander Graham Bell, a Scot, emigrated with his parents to Canada and thence on his own to Boston. It was there that, as well as teaching the deaf to speak, with financial backing he carried out his experiments and in 1876 achieved the first electrical transmission of intelligible speech. 'My God, it talks!' exclaimed the Emperor of Brazil on hearing Bell's primitive apparatus at the Philadelphia Centennial Exposition.

The technology soon came from the New World to the Old. Having married the deaf daughter of his backer, Bell came to England on his honeymoon in 1877 and gave many demonstrations of his telephone including one to Queen Victoria, who was presented with a £70 pair of instruments. The next year its possibilities were imagined at the Gaiety Theatre when Shakespearean and other actors played on *The Telephone Harp* and conveyed scenes from Stratford-upon-Avon to London. In 1879 the Telephone Company installed the country's first public exchange, at 36 Coleman Street in the City. These were hopeful beginnings. It seemed that distance need be no more a barrier to the transmission of speech than deafness.

But there were many who did not want to hear the message. In the UK neither officialdom nor the social system favoured the innovator. The telephone was to transform the Post Office for which Trollope had worked, but not yet. Its assistant engineer-in-chief, William Preece, recommended that the Post Office should secure the use of Bell's telephone and the right to manufacture. The opportunity was not grasped so private enterprise took the lead. Jealous of its monopoly on communication by electric signals, the Post Office claimed that the new instrument was a telegraph within the meaning of the Telegraph Act of 1869. The telephone operating companies rejoined that it was impossible an Act could include what was not merely unborn but unconceived at the time it was passed. They lost in the High Court. From 1881 the Post Office issued licences to companies to provide a service and undertook the general development of the system, without enthusiasm.

The government regarded the Post Office – the first nationalised industry – as a body for raising revenue through the penny post, savings, telegrams, dog licences, postal orders, and the inland parcels service. Establishing another telecommunications network meant the creation of an entirely new business, which involved investment in different buildings and equipment and the recruitment of a specialised staff. Such ideas would appeal to the few Fabians but this was not the function of government in a Free Trade economy. Even had Postmasters-General been willing they would have been restrained by the Treasury, exercising what Preece described as a 'stranglehold'.

The new means of communication was not to be developed as rapidly as had been the railways between 1830 and 1850. That change in physical communications transformed the speed and character of life and affected most classes in society. Telephone growth was to be much slower. Life was made brighter by dyestuffs, which founded the fortunes of the great European chemical companies, and by the incandescent lamp, which made possible domestic electric light and funded the development of other electrical apparatus. The telephone was not to be a great money spinner for its makers, typically engineers who worked for their City and Guilds qualifications but did not graduate to become members of the Establishment. Among them, it was not to produce any equivalents of the wealthy and influential press barons. Shrewd users of the instrument, such as stock and commodity brokers, were to make much more money from it, just as in the early years of the century Nathan Rothschild had helped to build his family fortunes by employing couriers and the pigeon post to speed intelligence.

Nor was the telephone to grow so quickly in importance in Britain as it did in the USA. There 'Ma Bell', than whom none would be greater or richer, was destined to become the supreme matriarch queening over the American economy, a fertility goddess speeding the application of ideas and hastening the tempo of business. The spreading use of the telephone greatly helped the growth of giant corporations and the formation of industrial trusts.

Their activities extended overseas, especially in new technologies that by their nature were international in scope. Western Electric, the company manufacturing telephone apparatus under licence from American Bell Telephone, was quick to extend its activities abroad. At first it operated through a European agent, Clark Muirhead and Company, but this British organisation did not match the international opportunities that had been created. Western Electric decided to handle the business itself and, in 1882, established the first factory of its kind in Europe, the Bell Telephone Manufacturing Company of Antwerp. In 1883, between the death of Marx and the birth of Keynes, it set up its own organisation in London. The two-man one-boy establishment was a small symptom of Britain's relative economic decline.

2

Advance, Mr Kingsbury: 1883–1898

Bell's father-in-law was angry when his own brother did not become the manager in London. Instead the job went to an unknown Englishman of 28, J. E. Kingsbury. His beginnings had been modest. When he was 11 his father's business as an auctioneer in Taunton had failed. The young John Edward had been chosen by a well-to-do uncle as the brightest member of the family of seven and brought to London to continue his education. On the journey up from Somerset the uncle travelled first class but the boy did not. He determined that when he grew up he would earn enough money to do so.

To the young boy London was a place of great promise, a city where he would be able to get on. Although he appreciated the English and Latin he was taught at Archbishop Tenison's School, he saw his way to success partly through his self-taught shorthand. This he practised by taking down sermons in church on Sundays, as he did not use it much during the working week in the office of his uncle, who ran Kingsbury & Co., an advertising agency in George Yard, Lombard Street. Just round the corner was one of the agency's clients, Colonel Gouraud, an American who was the London representative of Edison, then a rival of Bell. In 1878 the Colonel, needing some assistance with the correspondence on his telephone and electric light interests, asked the agency whether it could help and mentioned shorthand as a desirable attribute. Kingsbury was lent to him.

The secondment gave him some practical experience of the telephone. He helped put up a line between Gouraud's house and stables in Upper Norwood. Less successful was a cross-Channel trial in 1878, for which Kingsbury was entrusted with connecting one of the two Edison telephones in London.

The other was in the hands of a technical expert who had been sent over from the United States with all Europe as his province. As the business developed Kingsbury tried to work with him but 'he seemed desirous of maintaining the exclusiveness of his information'. To correct faults the young man had to conduct his own inquiries and research. 'If our expert happened to be in Paris and our instruments maintained a tactless silence I was generally appealed to for assistance, and earned something of a reputation as a magician.'

In his belief that 'there should be nothing mysterious about these instruments and that anything in the nature of exclusive information should be discouraged', he was glad to give lectures and practical demonstrations. He was familiar with the telephone in daily use in the offices of both his uncle and Gouraud, who were among the original ten subscribers to the Edison Company's demonstration exchange in 1879. Although he had no technical education to speak of and never laid claim to being an engineer, in 1887 he was to join the Society of Telegraph-Engineers and Electricians, the body that in the following year became the Institution of Electrical Engineers, of which he was to be a vice-president. Before publication, he would read his articles for the electrical and scientific press out loud, 'trying them on the dogs'. The telephone appealed to him early as something romantic. He could see a great future for it and thought that everybody should have one. A Gladstonian Liberal, he believed it would be brought to them through Free Trade, which he did his personal bit to promote. He would encourage his neighbours to use his telephone but would not accept payment. 'No,' he said, 'I'm a missionary for the telephone.'

In 1880 one of Kingsbury's brothers, Harry, was exhibiting in Australia the Edison electric pen, a stencil-making device manufactured by Western Electric. At the Sydney Exhibition he met F. R. Welles of Western Electric, who had been sent to introduce the company's telephone Down Under. Harry gave Welles an introduction to his brother John, which he duly presented on his way home to Chicago via London, where the Bell and Edison interests had merged into the United Telephone Company. Kingsbury and Welles were almost the same age and had similar temperaments: quiet, scholarly, to the world even austere. Both had started as personal assistants – Welles to Enos Barton, the president of Western Electric. They shared interests in literature and the telephone. Kingsbury gave Welles a copy of a shilling (5p) pamphlet he published in

1882 under the title *The telephone in principle and practice: its origin and development* by J. E. K. Fifty-three years later he commented: 'As a mere amateur I did not have the courage to put my name in full.'

The personal and business relationships grew together. Welles talked in general terms about opening a London office now that he had established the Antwerp factory. Having satisfied himself that the idea was sound, he acted quickly. Within a few weeks, on 2 May 1883, an office and small store were opened for business at 59 Moorgate with Kingsbury in charge. Most of such appointments went to men with telegraphic experience but Kingsbury knew the telephone and several people with whom business could be done. The great advantage he had over the brother of Bell's father-in-law though was that he was an Englishman. He had become a Londoner. Welles, much travelled and a linguist, believed in appointing a native to control local business and giving him the widest freedom of action within the principles of the company's policy.

Initially Kingsbury had less room to manoeuvre in the market. The first licences granted by the Post Office in 1881 restricted the area of telephone communication to a radius of five miles from a defined centre. Gladstone's Postmaster-General, Henry Fawcett, encouraged the formation of private operating companies – the Post Office got a ten per cent royalty on their business – as 'it would not be in the interest of the public to create a monopoly in relation to the supply of telephonic communication'. In 1884 the five-mile restriction was withdrawn but the licences were retroactively to run for a maximum period of thirty-one years, ending on 31 December 1911. They could be terminated in 1890, 1897 or 1904.

The policy was not exactly a stimulus to development and growth. There were some local networks but they were not linked by trunk lines. In 1887 a Conservative Postmaster-General told Parliament that 'having regard to the cheap and swift means of communication which at present exist by means of the telegraph between the principal towns in the UK . . . it is extremely doubtful whether there would be much public advantage in establishing telephonic communication generally between those towns'. He was anxious about the threat to his telegraph revenue.

The only place it was possible to telephone from London was Brighton. Not surprisingly, the opening of two circuits between London and Paris in 1891 led to demands in Parliament for the establishment of a proper internal trunk network. By the Telegraph Act of 1892 the Post Office was given £1 million to buy the trunk lines from the operating companies and create a national network of high standard, to which all subscribers would have access. This was substantially complete by 1896, twenty years after Bell's invention. The quality of reception might leave something to be desired but at least there were lines from Aberdeen to Plymouth. Demand, however, for additional trunk lines was growing.

Britain was behind many other European countries in practical developments and public attitudes. Conan Doyle gave Sherlock Holmes a telephone at 221B Baker Street by mid-1898 but the great detective conducted most of his urgent business by telegram, a slower and more involved means of communication that gave him less privacy. His problem would have been that the people he wanted to reach were not on the telephone. Original expectations were that the principal business firms would become exchange subscribers and it was not contemplated that the service would become general. Although the telephone received royal patronage, possession of one was not a mark of social status. Oscar Wilde found no need for it as a dramatic device in his comedies of manners and most households managed well without it. They were more likely to have a servant, who for a similar cost, about £20 a year, did much more. Even in Kingsbury's own house the maid was terrified of the telephone up on the wall, believing it to be operated by some sort of witchcraft. When it rang she would not look at it and, if an adult were not present, she would lift up one of the children to answer it.

Looking back at British attitudes, Kingsbury wrote in 1916:

> The service grew as its utility was demonstrated, and grew not merely without external assistance, but in spite of every possible difficulty put in its way. Governments, both local and national, interposed restrictions; private property owners desired its use for their own purposes, but refused any assistance for their neighbours. Having no legal powers or public way facilities, the companies were dependent upon wayleaves over private property, which were difficult to obtain and very expensive to hold.

Wires were unsightly, especially in clusters near exchanges, but the objectors were not altruistically protesting about the environment.

The obstacles did not discourage the young man when in

1883 he started the business, which was in competition with several electrical engineers trying to apply the new technology. As well as his missionary enthusiasm for the telephone he had an incentive to succeed. He wanted to marry Kate, which he was happily able to do the following year on his initial salary of £300. Business was good and his salary went up to £360 in 1885 and to £500 the next year. Not that his guiding motive was personal gain. Good businessman though he was, he saw himself more as performing a professional service than as being in trade. Orders were placed by the operating companies – most of them had merged into the National Telephone Company by 1889 – and, in a smaller way, by the Post Office. Kingsbury himself wrote the orders on a piece of notepaper, which he mailed to the overseas factories with a covering note, perhaps incorporating hand-drawn sketches of parts, explaining the use the customer intended to make of the apparatus. He was meticulous in keeping exact accounts so that the parent company knew the precise trading position.

Finished equipment was imported from Western Electric in Chicago or Bell Telephone in Antwerp. A Belgian minute of mid-1891 notes 'the orders for switchboards and telephones greatly exceed the means of production. These orders originate chiefly in England.' The new technology prospered in a depressed economy. In need of more space, the agency had moved to 79 Coleman Street and taken a store at 13 Cross Street, Finsbury. In April 1894 Frank Welles wrote to Enos Barton, the president of Western Electric, from Alassio on the Italian Riviera, where he was resting from the pressures of business:

> After being in Antwerp awhile, from the beginning of February, I became conscious of the fact that the factory was running almost entirely on London orders, the Continental trade being very small indeed. As the London orders are almost entirely from the National Telephone Co. we have the very unhealthy situation of both London and Antwerp being dependent for their existence on a single customer.

Up to that time Western Electric had been able to command good prices for its equipment because American enthusiasm for improvement led to technical advances that did well against the competition in international markets. For example, an early problem arose in the USA with the growth in the number of users. Connecting any two subscribers in an exchange was not difficult when there were only a few of them. One operator could handle a group of up to fifty, but above that number more operators were needed. Simple switchboards necessitated operators – in the early years they were boys – calling to each other across the room to make the required connection. The more subscribers, the greater the problem. To overcome it, each operator had to be provided with facilities for connecting any one of his or her lines with any other line in the exchange, but not when a line was already engaged. The answer was the multiple switchboard, the first British installation of which was in the dome of the Liverpool Cotton Exchange in 1884. It enormously increased the number of lines that could be handled on a single exchange but the cautious Liverpool directors stipulated that it should be taken out again after three months at Western Electric's expense if it was not satisfactory. In fact 1,000 subscribers were dealt with by ten operators, with an average connection time of fourteen seconds. Designs improved but the principles held good.

Liverpool was also the site for the installation of the first dry core cable in England. Laid through the Mersey Tunnel in 1891 to connect Liverpool and Birkenhead, it was a general technical advance that others too were to exploit. Samuel Morse's original telegraph line had failed because the cotton covered wires were not an effective insulation. Cotton, paper and other available fibres absorbed moisture, which caused electrical loss. They could only be used successfully as insulation materials if they were installed and maintained in an artificial state of dryness. The early solution was oil filled cable.

W. R. Patterson, a Chicago cable engineer, saw the problem:

> Telephone exchanges were developing to such an extent that underground cables in city streets had to be considered as well as aerial cables, to avoid the congestion of wires leading into offices. The Brooks cable was not suitable for such uses; it was very expensive to construct, as all the work was done in the field, at the mercy of the weather, and pipes from which kerosene oil was intended to leak were not looked upon with much favor in streets and buildings.

Patterson was convinced that there was no better insulation than air 'but the people who put their good money into cables were not convinced; it would have been as easy to sell a cable filled with air as to sell a sausage filled with air!' It took him several years and nearly 100 patents to arrive at a practical process based on charging melted paraffin with a gas and allowing it to cool under pressure. Later he commented: 'We all did a lot of fool things in the early days of the cable business. Still, it was all pioneer work. We had no precedents except failures, and not many even of those.'

Over thirty years after the Mersey Tunnel installation Frank Gill, who had been chief electrician to the Liverpool District of the National Telephone Company, pointed out that the method of jointing the cables in the manholes was almost identical with present methods. In those days such techniques were designed and carried out on the spot.

During the early 1890s improvements were also introduced in the design of the carbon microphones used in telephones. The self-restoring indicator, which preceded lamp signalling, was first used on exchange switchboards in Hull and then in larger exchanges in London. By the time of Queen Victoria's Jubilee in 1897 telephone systems were robust enough to go into temporary service at short notice. For the army review and torchlight tattoo at Aldershot on August Bank Holiday, Kingsbury was asked whether telephones could be installed to facilitate troop movements. Nobody had done it before but he was confident enough to depute Frank Jewson, barely two months with the company, to look after it:

> On the Thursday preceding the Bank Holiday I was instructed to select from the Stores whatever material would be required, and to proceed the next morning to Aldershot to install the equipment, which should be in operation for the final rehearsal on the Saturday afternoon. The equipment selected consisted of a 50-line standard board, 40 magneto wall sets, and supplies of batteries, wire, solder, etc. A company of sappers and miners ran the necessary line wires to a suitable spot where the Switchboard could be located out of sight but near enough to enable the operator, a corporal, to receive verbal instructions from the officer in charge of the manoeuvres. By working right through the Friday night everything was ready for the rehearsal on Saturday afternoon, and after seeing that it operated satisfactorily I returned to Waterloo Station in an otherwise empty troop train in the small hours of Sunday morning.

One of Jewson's normal jobs was to accompany Kingsbury to the National Telephone Company when it wished to place an order. From the discussion of requirements he wrote a specification and prepared floor plans to be sent to the factory. His assistance was a great help to Kingsbury, who had been feeling the strain of working without adequate support. Jewson was also responsible for testing imported apparatus and making any necessary repairs or adjustments before delivery to a customer. That was not difficult with small apparatus but drums of cable, then made in Paris, could present a problem in the small stores. Manufacture was also too far from the market.

There were other pressures on the company to manufacture in the UK. Its biggest customer, the National, had threatened to start its own factory, partly to avoid buying foreign products. Overseas, as industrialisation proceeded, tariffs were raised against British goods, while at home, after half a century of Free Trade, protection was being talked of again. A member of the National Liberal Club, Kingsbury was disappointed in the trend but he realised that practically and politically it would pay Western Electric to establish its own factory. One of the sites he looked at was on the South Bank of the Thames where County Hall was later built. The costs and conditions of leasing land and erecting a new building across from the Houses of Parliament were, however, unacceptable to the Western Electric board.

Frank Welles, who visited England monthly, had been keeping his eye down river on the uncertain fortunes of the Fowler-Waring Cables Company at North Woolwich. Registered in 1889 to take over certain American patents in the UK for improvements in the manufacture and insulation of lead-covered electrical cables, it made a profit in 1892–3 because the directors waived their fees, and was reconstructed in 1895. It still did not prosper, as Welles noted in a letter to Barton on 29 March 1897:

> Had a pleasant call on Mr Preece [now Post Office engineer-in-chief] last time I was in London. We told him and everybody of our intentions as to factory, which evidently made a good impression.
>
> The F. W. Cables Co. are promised £30,000 additional capital by the Fowler of the concern. They offered our Mr Turney £600 a year, with increase to come shortly, to take the management of their business. He talked it over with his wife, and decided that he was safer with the W. E. Co., at less than half the pay.

On 4 August 1897 the National placed its first large-scale

contract with Fowler-Waring for the supply of dry core cable containing 61,200 miles of wire. The company was to fulfil each ordinary order at the rate of three miles of cable per week and each emergency order at the rate of six miles a week, all to a tight specification. Profit expected on the contract was £15,000 – but all was not well. Negotiations were proceeding for Western Electric to buy the company.

The very next day Welles sent a hand-written letter to Barton in Chicago giving Fowler-Waring's valuation of itself at £95,000, including £20,000 goodwill, most of which was made up of the expected profit on the National contract. In his reply of 25 August Barton resisted allowing anything for goodwill, recognised that it was a 'question of taking on a permanent investment approaching half a million dollars' and that 'negotiations with an English corporation are likely to be slow'. Welles had the assurance that if he could work out a scheme that he could personally heartily recommend as satisfactory it was pretty likely to go through.

In the event the deal was done quickly, although later Western Electric went to arbitration over the stock. On 24 November, meeting in Boston, the Western Electric board unanimously agreed to buy the freehold property and business of Fowler-Waring for £87,000. The effective date was 1 January 1898. It was a precedent for profiting from another company's misfortunes, but in the meantime it added to Kingsbury's problems. The busy agent had to take over an unwell company, bring the plant up to American production standards, and guide the enterprise into the twentieth century. All this added up to a much larger and very different managerial role. Circumstances were not to be on his side.

3

Making and Breaking: 1898–1914

The company's first plant was on the stretch of the Thames where many telegraph cables were made. Cable-making being similar to rope manufacture, the original cable-makers were wire rope companies. They had the equipment for drawing copper rod into wire and twisting it into conductors. Insulated with paper or cotton, the conductors were put together in pairs to carry a two-way conversation, numbers of pairs forming the core of a cable. The dry core was then sheathed with lead against moisture, and for added protection it might be armoured with steel wire.

In cross section the design of a cable is simple, but like a rope it is only as good as its weakest point. The skill lies in manufacturing and testing processes to produce a cable consistent in construction and, over a long period, reliable in performance. The North Woolwich plant was well placed to do that economically. Heavy lead, copper, paper, timber and coal were brought by barge to be unloaded on the 325-foot wharf, and finished products could be transported by rail, road or river. The shop floor skills had long been in the area, in places like W. T. Henley's Telegraph Works next door, Siemens Brothers, and the Telegraph Construction and Maintenance Company. North Woolwich was an East End village, an artificial peninsula bounded by docks on the north and the river on the south. It was a natural community and 'The Western' became one of the obvious places to work. Those who lived across the river could come to work on the free ferry. Kingsbury did, cycling from his home in Blackheath. When fog stopped the ferry running, employees came over, eight at a time, in a rowing boat.

Business got off to a good start, as Frank Welles explained in a letter of 31 October 1898 to Harry B. Thayer, a go-getting

New Englander who had doubts about Kingsbury's methods:

> It would certainly be difficult for K. to get away at present for any length of time; the whole concern in London rests on his shoulders. I noticed it particularly when there last week; he has to attend to almost every little thing, and the business is now a totally different affair from what it was last year, the taking over of the F.W. factory having made a revolution in it.
>
> K. kept about all the F.W. office people, only dropping the old manager, and as you know we decided to move down the City offices and staff to N. Woolwich. The move was made, and is a success, but it was found that the arrangement of the factory offices was bad, and we are adding a storey to a small office building so as to get the people all together.
>
> You can well understand that the taking over of a factory and business, with many things to be revolutionised, and the two sets of people to be fused together into a harmonious whole, has involved a lot of work and worry. Besides, the settlement of the F.W. bargain has eaten up no end of time and is not yet over. We went to arbitration over the stock, and I was called on while there to give evidence, being examined by Kingsbury and cross-examined by the former F.W. manager. When that row is over there will be more time to attend to business.
>
> Of course the present condition of things is not what it should be, and K. is doing his best to get the people organised into workable shape, so the machine will stand alone and run itself. But that is a process that takes time . . .

Transatlantic discussion of a manager's abilities and attitudes was a recurring feature of life in the company, the degree of misunderstanding depending more upon the people involved than the communications facilities available. In writing to Barton on 1 May 1900, the normally sympathetic Welles commented: 'Kingsbury's obstinacy in sticking to an unsuccessful policy or an unsuccessful man is a drawback in London'. Apart from his support for local staff, one of the issues on which there may have been a difference of opinion was joining the newly formed Cable Makers' Association. A protective club like this went against his Free Trade principles, even though becoming a member may have stood him in good stead locally and internationally. Welles recognised that, among the senior people in Europe, Kingsbury 'is in more need of money than anybody else'. He was then being paid a London salary of £600 and £120 as a director of the Antwerp company.

The National business had certainly come on. In a contract of 22 December 1898 Western undertook to supply between 1 January 1899 and 1 January 1905:

(a) A yearly minimum quantity of 12,000 telephone wall sets of the newly-adopted type during each of the first three years at an agreed price of 42s [£2.10] each.
(b) A yearly minimum quantity of 12,000 micro telephones of a type to be approved during each of the first three years at an agreed price of 14s [70p] each.
(c) All the new exchange switchboards for over 600 subscribers that the Company may from time to time require.

Under a contract of 13 April 1899 Western was to supply dry core cable containing 30,600 miles of wire, with a minimum rate of delivery of 1,632 miles of wire per week. All went well until 21 July 1899, when as the *Stratford Express* reported:

> A very disastrous fire broke out early on Friday at North Woolwich in the large factory of the Western Electric Cable Company, and a three storey workshop measuring 250 feet by 70 feet containing valuable machinery of the latest pattern was completely destroyed, the damage being estimated at £50,000. The workshop was closed on Thursday evening at 6 o'clock. The watchman noticed nothing to raise the slightest uneasiness until about half past one in the morning when he observed an unusual light in one of the many windows and found that a fire was raging in the top floor. A dozen men who were on the night shift got out the fire appliances belonging to the factory and endeavoured to check the spread of the flames while a message was sent to the North Woolwich fire station.
>
> Fire engines from Silvertown, Blackheath, Deptford and other places soon arrived. The alarm was first given to the West Ham brigade at 1.30 a.m. the message being conveyed from the Silvertown pumping station. All the engines were called out and East Ham was also sent for, while Metropolitan engines arrived in large numbers. Altogether 18 steamers, 1 float and 2 hydrants were brought to the scene and 13 steamers and the float were got to work but the factory and stores in which the outbreak was discovered were totally destroyed and an adjoining corrugated iron building was damaged by water and breakage and some goods in the yard damaged by fire . . . . We are informed by the Western Electric Company that the instrument department is not affected by the fire. By making temporary arrangements they will be able to execute the orders of the majority of their customers with but little delay and the business will be carried on without interruption.

The company was able to fall back on the help of associate companies, but the disaster caused Kingsbury months of personal strain. He was aware that for years it had been the policy of the National, sensitive to criticism about its volume of imports, to get free of Western Electric. What kept them together was Western's technological progress, seen at first hand later that year when the general manager and chief engineer of the National visited Chicago and were impressed by the common battery switchboard. It was a piece of equipment they wanted to install in their important exchanges as quickly as possible and it made them reverse their policy towards the company. Their decision did not help Kingsbury, however, who had to wait for a better plant to rise from the ashes.

There must have been serious transatlantic misgivings about the future of the enterprise because it was not until November 1901, over two years after the fire, that the board authorised $300,000 (£61,500) for reconstruction. The cable plant was redesigned and rebuilt by 1904 and a telephone instrument factory was then added, at first using imported parts. Catalogue prices of wall-mounted sets ranged from £3 to £4 10s (£4. 50). By 1908 complete switchboards were being produced, the National and the Post Office ordering two different types, with only one relay being common to both. The Post Office switchboard was lower, in mahogany, with the cable painted in dark asbestos – fires were a real hazard – whereas the National preferred light oak and light grey asbestos paint. Matching the polish in an extension to equipment the customer had kept in a high shine could be a devil. It was the kind of detail the bowler-hatted foremen kept an eye on.

Expansion into this kind of manufacture involved further building and establishing new departments, from a drawing office to a metal-finishing shop. Outside the general office on a bench in a corner a man, his son and two girls made condensers out of paper and tinfoil. It was the beginning of what was to become a big business in an expanding range of electronic components, each type initially being made for a new or improved product. Installation and engineering departments for cable and apparatus had to be built up. Although some of the clerks still sat on high stools at Dickensian shuttered desks and wrote letters by hand, which were copied in a wet press, a small typing pool produced the more important correspondence to customers. By 1909 the company employed almost 1,000 people. There was a variety of jobs in a growing industry, whether one wore a leather apron, a white linen coat or a dark suit. But wages were low. A man repairing cable drums got 5¾d (less than 2½p) an hour for a fifty-two and a half-hour week and lads 2¼d (less than 1p), lasses 2d.

In skilled jobs people could contemplate a progressive career. By 1912 there were five-year apprenticeship schemes, with schooling being done at Woolwich Polytechnic. The plant was of sufficient substance to attract young Englishmen of promise, people who were to be important in the company's management. George Howard Nash, who had been apprenticed to the National, was a future chief engineer. In the drawing office, recruited from Woolwich Arsenal was Thomas Spencer, who was to be the company's longest serving managing director, and beside him John Pheazey, a future works director.

There were technical developments to attract them and hold their interest. In working for an East End company they were getting continuous access to the latest technology from America, the source of most new inventions. In 1899 the UK's first common battery switchboard, made in Chicago, had been demonstrated in the Woolwich works and the first installation for public use in Europe was put into service at Bristol for the National. The essential feature of this was the elimination of batteries from the subscriber's telephone by the supply of a uniform current from a large central battery at the exchange. Introducing the use of electrical power to the telephone system, it was a stimulus to the development of relays and other electromechanical devices to extend automatic signalling and supervision in exchange working. One result was that an operator could watch a conversation on small electric glow lamps and did not need to butt in with 'Have you finished yet?' The subscriber also needed simpler, cheaper apparatus and no longer had to mess about changing the sal ammoniac in a Leclanché cell. The new arrangement led, when unauthorised instruments were connected, to the curious charge of stealing Post Office electricity.

In 1906 the first coin-operated call box for the Post Office was installed at Ludgate Circus. From 1908 the distance over which clear and audible calls could be transmitted was increased by the addition of loading coils to circuits. The mathematical theory behind these had been propounded by the English physicist Oliver Heaviside as long ago as 1887 but it was Pupin, an Austro-Hungarian immigrant to America, who devised a practical application, and Western Electric that put it into production. Manufacture of the coils was added to the work at Woolwich and from it was later to develop another component activity, magnetic materials. After successful installations in the metropolitan area, a notable contract was for the first Post Office loaded trunk cable between London and Birmingham. By the use of loading coils the diameter of copper wire in a cable could be halved, a significant saving when the metal accounted for about a quarter of the capital investment in the telephone system. It was the commodity price that Kingsbury checked daily in the *Financial Times* because it also affected contracts for power cables, like those the company carried out for the electrification of the Bakerloo line on the Underground, for Brighton tramways and for installation below the Blackwall Tunnel under the Thames.

During the Edwardian period work for the Post Office was increasing. The attempts at municipal telephony between 1899 and 1906 were largely a failure, with only Hull destined to survive. Formed in 1903, this local authority enterprise was an early and continuing customer of the company, which had learned from its lack of success in tendering to similar authorities. In practice, the National was confined to its existing areas of operation, where it modernised by replacing some magneto exchanges by common battery equipments. Under the 1905 agreement with the Post Office it could look forward only to its plant and business being taken over on 31 December 1911, with no payment being made for goodwill or profits. The National therefore had little incentive to invest, which must have meant that during its period of rebuilding, Western got less business from its major customer. Between the 1902 order for the 12,800-line central battery exchange at London Wall, which was manufactured in Antwerp, and the 1906 14,400-line Hop exchange installed in South London, there seem to have been few significant equipment orders.

Cable business was probably also not what it might have been. A founder member of the mains section of the Cable Makers' Association in 1899, the company resigned in 1903 and there were protracted negotiations over the terms of rejoining. At issue was some sort of guarantee on either £15,000 of rubber cable work, a modest allocation, or a fair rate of profit on an equivalent amount of business. It seems that on the home market the company was losing out to larger competitors like Callenders, Henley's and Siemens. For

Tunbridge Wells in 1900 it put in the highest tender for supplying and laying armoured cables, over half as expensive again as its successful neighbour, Henley's. In business Kingsbury probably had nothing like the Liberal landslide of 1906 to elate him and he had more than one reason to dislike the impending Post Office monopoly.

To some extent restricted growth at home was made up for by business overseas. To add to the office opened in Sydney in 1895, branches were established in Johannesburg and Buenos Aires and agents appointed in India, Egypt, Portugal, the Straits Settlements and other countries then in the territory of London. Work was undertaken in places as far apart as Rumania and the West Indies. Up to 1906, however, the return on sales and investment was poor, below what the parent company expected. Between 1900 and 1909 Kingsbury's salary on the London books was static at £600 and his total emoluments, taking into account his director's fees and additional remuneration on the Antwerp books, grew from £720 to only £1,020.

The precise Harry B. Thayer, who succeeded Barton as president of Western Electric in 1908, put his point on Kingsbury and 'the London situation' – not the last time this all-embracing phrase was to be used across the Atlantic – to Welles in a letter of 25 January 1909:

> The thing that he can do better than anyone else is to conduct the high line of policy and negotiation with the Post Office and the National Telephone Co. The thing that someone else can do better than he is to manage the business.
>
> I am satisfied that his work ought to be limited to relations with the Post Office and National Telephone Co. and properly, it ought to be subordinate to the North Woolwich management.

Welles, who preferred the existing arrangement, gave a copy of the letter to Kingsbury, who wrote to him on 4 March 1909:

> It is in accord with the line of development followed in the States . . . I believe the Company's English interests require more Anglicising than Americanising.
>
> I hold more strongly now than I ever did that the full development of the Company's business here requires a directing force of local origin which shall influence the general tone and policy of the Company's operations as a whole. Mr Thayer holds in effect the contrary.

In such a matter of principle New York was bound to win, especially as it intended to establish a British company. After an interchange of carefully worded correspondence Kingsbury stepped down and his place was taken by Irving Loveridge, a self-contained American who would walk through the office looking straight ahead and ignoring everybody. He was in office for only a few months and his untimely death delayed the incorporation of the company. He was succeeded by another American with a financial background, the Western Electric special agent in China, G. E. Pingree, who opened and read all the company mail and came into the office in his white spats on bank holidays as well. In the immediate future the operating style of the company was to be one of overt dedication to the business.

On 10 January 1910 the Western Electric Company Limited was incorporated with a capital of £500,000 and acquired for £189,000 the business and property in the UK of its parent. Control was firmly with the parent, the minor decisions being taken in the UK head office at Norfolk House, Norfolk Street on the Embankment. The policy was clearly stated:

> Our foreign houses are largely managed and the shops supervised by men appointed from the home organisation – men who are familiar with our methods, who are in sympathy with our aims and desires, and who understand the company's viewpoint in regard to its foreign work.

In theory, plants primarily served their own markets but in practice they specialised and could not meet all the needs of those markets. Thus Woolwich was primarily a cable plant and still had to import much of its telephone equipment. With the rise of locally owned competitors in manufacturing, it was more important than ever to be on good terms with the telecommunications operating bodies.

This may have been a factor in bringing about a rapprochement between Kingsbury and New York, where he had a number of friends. At 55, after serving for twenty-six years and nine months, he was granted a pension of $103. 45 (£21. 40) per month on the personal recommendation of Thayer, invited over to discuss his future, and made a director of the new UK company. His compensation was fixed at the same amount as he had received as manager, $4,929. 23 (£1,020), made up of an annual fee of $3,687. 83 and pension of $1,241. 40. Relieved of day-to-day responsibilities, in the company's annual return he was described as 'gentleman'. In fact, much of his time was to be spent in writing *The Telephone and Telephone Exchanges*, a comprehensive study eventually published in 1915 and destined to remain a standard work. The word 'gentleman' was also used of H. M. Pease, a Southerner who became sales manager and eventually more English than the English. Not all those from the USA who occupied the key jobs could be so described. One of them kept a spittoon at some distance from his chair. Another had eyes that looked right through people. Some were flashy in their manner. As George Bernard Shaw had found in his short period with the Edison Telephone Company, the energy they expended was sometimes out of proportion with the result, desired or actual. Doubtless they were quick to claim credit in Chicago for changed market conditions in Britain. Their aim was to handle a bigger volume of business more efficiently by making North Woolwich more of an American outpost.

In January 1910 the site area was doubled with the acquisition for £22,500 of the land that Welles had long had his eye on, four and a half adjacent acres belonging to the Telegraph Construction and Maintenance Company. Even though this could not be fully developed until 1913, Western Electric wanted to be prepared for the extra business anticipated when the Post Office formally took over the telephone service from 1912. Like the formation of the Port of London Authority and the amalgamation of the London private transport companies, it was expected to lead to an extended and improved service. One body would avoid duplication of effort and waste of resources; it would have a clear policy of development. The Post Office would henceforth be the company's major customer.

At the time of the takeover most of the country's exchanges belonged to the National. Like those of the Post Office, all were manually operated. Yet the first patent application for an automatic system by Almon B. Strowger, the Kansas City undertaker who lost business through calls being connected to rivals by an operator, dated from 1889. To its credit, immediately it became a monopoly the Post Office set out to investigate and get practical experience of different automatic systems. On 18 May 1912 the first experimental Strowger system, made in Chicago by Automatic Electric Inc. and installed by the Automatic Telephone Manufacturing Company of Liverpool, was put into service at Epsom. Another 500-line exchange, known as the Lorimer system, was imported from Canada and put into operation in Hereford on 1 August 1914.

Western Electric made its contribution with an Antwerp-made 2,800-line rotary exchange cut over in Darlington on 10 October 1914, by which time Europe was at war. Orders secured by London for similar exchanges to be installed in South Africa and New Zealand could not be immediately fulfilled.

At the time there were in the UK only 1·7 telephones per 100 of the population, against 9·7 in the USA, 6·5 in Canada, 2·8 in Australia, 3·5 in Hawaii and 4·6 in New Zealand. These comparisons created for Western Electric both opportunities and impatience. Besides low telephone density the UK had no system worthy of the name or the times. The Post Office was beginning to feel its way as the monopoly operator. With its research facilities consisting of four overcrowded rooms at Dollis Hill in North London, it relied largely on foreign manufacturers for its expertise. Kingsbury, a missionary to a stiff-necked people, was not harsh in his verdict that 'the facilities offered by the telephone have been but poorly utilised in the UK'. Almost forty years after its invention, civil progress in its application was to be delayed for another five years by war. During that time America would continue to advance.

4

The Business of War: 1914–1918

On 3 August 1914, the day that Germany invaded Belgium, Antwerp was declared to be in a state of siege. It was decided to transfer Bell Telephone Company's central engineering and automatic exchange departments to London 'for the time being' and evacuation of the English and most of the American personnel began the following day. North Woolwich had to adapt quickly to wartime conditions, as cablegram reports to New York show:

> 5 October: London conditions normal. Shop active. New orders coming in good volume which will keep some departments busy until March.
> 20 October: London very busy both apparatus and cable shops. Personnel maintained replacing those joining colors. Using good advantage Antwerp refugees.

It was not entirely one happy international family. There was some resentment at the easy life refugees were having, especially when they relaxed in, say, a game of cards. English employees were not slow to murmur that their sons had gone to war for Belgium and were dying in Flanders fields. One of the immediate effects of the war was to dissolve the 1913 agreement between Siemens & Halske and Western Electric, whereby technical information was exchanged and business shared to avoid cutting one another's throats. Hatred of the Germans was rife. The girls in the cable shop threatened to strike if Americans with German-sounding names were not pushed out. They were repatriated within a week. On 24 July 1915, 468 Chicago employees lost their lives on the annual excursion and picnic when the steamer *Eastland* overturned in dock. Seeing names like Schneider and Schultz among the death roll in the mauve-bordered issue of *Western Electric News*, one Woolwich employee callously remarked: 'This almost sounds like a victory over the Germans, doesn't it?'

In the government-controlled plant, under mainly American management, most of the effort was directed towards the war in various ways. There was little export, with some small jobs being done for the United River Plate Telephone Company. Production and installation of civil telephone equipment slowed down, except where it was essential for the war effort, as in munitions factories and expanded ministries temporarily housed in mansions and hotels. One such installation later in the war was for the American naval headquarters in Grosvenor Gardens. Completion in 1915 of the London to Birmingham loaded trunk cable, the longest of its kind in Europe, brought much better communications to the home of the small armaments industry. Manufacture of munitions at the Woolwich plant was restricted to the lighter types that could readily be produced with the machines and processes normally used for making telephones: conventional items like hand grenades, rifle grenades, primers, fuses, detonator caps, and parts for mines, cordite drums and cable reels, which in the field were carried on the backs of horses or men.

The telephone, first used on active service on the North West Frontier in India in 1877 and in the field in 1882, had been increasingly applied as a basic weapon by various armies, most recently in the Balkan Wars (1912–13). In the more industrialised First World War it provided communications between the firing line and the successive headquarters in the rear. Scouts with portable equipment were able to report back before returning and the faster information enabled artillery fire to be better directed; but on the Western Front it did little to alter the slogging character of the to-and-fro trench and hill warfare. On the Eastern Front field wire and cable were in demand for telephone and telegraph purposes, and payment for a quantity supplied to the Russian Government was still being sought ten years after the 1917 Revolution. Signalling equipment made at Woolwich for the Serbian Army included single- and four-line portable sets with bought-in accessories like insulators, poles and earth rods. Work on a single-line set for the Russian Army, ordered in such quantities that it had to be produced in Chicago, led to the American adoption of some of the English craftsmen's standards in the production of leather cases.

In more significant ways the stimulus of the war and the comparative isolation from Chicago – Americans were less keen on transatlantic travel after over 100 of their countrymen lost their lives in the *Lusitania* – encouraged native talent. A twelve-line portable switchboard was developed and packaged in a khaki wooden case for field use. Also developed was an airship telephone system including throat transmitters for the use of crews in 'Blimps' that patrolled in search of enemy submarines. Some work was done for the Royal Naval Air Service on an early guided weapon system for directing an aerial torpedo towards a source of sound. The work was abandoned because of the success of aeroplanes in dealing with Zeppelins but it was important for two reasons. Technically it involved the first use of valve amplifiers, specially imported from America and destined to be of postwar importance in improving long-distance lines. Immediately, it was another boost to the confidence of the home-bred engineers, whose role hitherto had been in carefully following transatlantic technology in civil communications. The exigencies of wartime now demanded that they come up with their own ideas and turn them into practical hardware quickly.

The chief engineer Nash, a handsome, flamboyant man who aped his American colleagues, especially in quick-fire decisions, was not above taking credit for other people's work. In 1915 it was discovered that a German listening post was eavesdropping on telephone and telegraph messages being sent to the British front, amplifying the weak signals and making them audible. To jam the enemy's reception Nash and his team devised and demonstrated to the War Office a screening set that created a continuous buzzing. Sets were successfully used at a time when front line information in the stalemate of trench warfare was valuable.

Another necessity in those conditions was some means of detecting and locating enemy mining operations in no man's land. If it was possible to record an earthquake at a distance then surely tunnelling only yards away could be identified. On a similar principle to the seismograph the Woolwich engineers produced the seismophone, which was based on the carbon microphone used for transmitting speech on the telephone. Like the mechanical vibrations picked up from the human voice, the movements of enemy sappers were converted into electrical signals. Means for determining distance were added and the assembly packed in a heavy-duty shell-shaped case connected by wires to a box upon which the operator sat with his headset. The receiver was calibrated for different types of soil and exact enemy locations were found by triangulation. Four-detector units were produced for the Western Front and simpler models with only one detector for Russia and Italy.

Aware of the achievements on land, the Admiralty wanted something similar that could detect enemy submarines. From 1916 U-boat warfare had intensified as the success of the British naval blockade had led in Germany to winter food riots, strikes and a rising infant mortality. In the second quarter of 1917, when the USA came into the war in response to the German policy of unrestricted naval warfare, British merchant shipping losses peaked at over a million tons. The race was on to produce an effective deterrent. By July, within the three months specified by the Admiralty, Western Electric was able to demonstrate convincingly in Weymouth Bay that its experimental hydrophone could do the job. Its first unit for active service was ready by 17 October 1917.

It worked on the same principle as the mining detector. The specially-designed listening devices were mounted in a balanced torpedo-shaped body, which was towed at a given distance below the surface. Not surprisingly, the streamlined unit in gleaming nickel silver, its hundreds of screws shining, soon became known as the Nash Fish, a term that the immaculately dressed Nash did little to discourage. Many believed that most of the work had been done by one of his assistants, R. A. Mack.

On board ship were a valve amplifier in a stout teak case, a mahogany control desk connected by telephone to the bridge, and depth charges. A training school was established at Portland and there was a rush in 1918 to install the necessary equipments, twenty skilled fitters being temporarily transferred from the Post Office to supplement the company efforts. Most of the installations, over 200 in all, were in trawlers, with a few on destroyers. Three patrols of submarine hunters were established: to the north east of the Shetlands, off Plymouth, and in the Adriatic. They made some kills and probably had a greater effect on the morale of enemy crews. The hydrophone performed well enough for fifty equipments to be ordered just before the Armistice, but there is no evidence that it had a significant impact on the final outcome. The parent company never took up manufacture as it did of other war equipment.

Production was occasionally interrupted by Zeppelins

passing over the Woolwich plant but the only damage was done by a stray anti-aircraft shell that fell in January 1918 without exploding. There was more concern over subsidence on the site, noticed in the middle of the war and estimated to cost more than £20,000 to repair. The biggest changes were human, as G. C. Goodburn, the employment manager, described in *Western Electric News*, his English having been sub-edited:

> The force of female workers of the London House has increased 100 per cent since 1914. Some have forsaken the shop counters of grocery establishments or drug stores, some from the millinery and dressmaking professions, and others from domestic service have come to us to enter the more important duties connected with the manufacturing departments of a factory, often necessitating rising in the very early hours for the purpose of clocking on at 7 a.m.

At 6.55 a.m. the hooter went and employees lined up outside Number 5 gate were allowed to clock in. The gates were closed at 7.00 precisely and opened again at 7.03, by which time anybody clocking in was docked a quarter of an hour's pay.

Goodburn went on:

> Where in 1914 you saw Jack Jones you now see Cissy Brown at the capstan lathe in the screw machine and fuse departments. Similar changes may be found in the personnel in charge of the various lifts, and porteresses in their smocks in the stores departments truck the scrap from the machine shops to the scrap bins. Girls work on ledgers in the accounting department. Some make cable reels and cases, some are on inspection work and others on work which, but a short time since, would have been deemed impossible by some and unmaidenlike by others.

North Woolwich was not the most respectable of places in which to work. On the line from Stratford, for example, there were unsavoury characters like cardsharpers going down to the docks and women ready to claim their virtue had been outraged in a non-corridor compartment. Working at the Western though was considered a cut above being at Tate's sugar factory or Keiller's jam works, where the girls wore shawls and wooden clogs and were said to prey upon unsuspecting men. The advertising manager described the plant as 'one of the finest and most up-to-date plants for the production of telephones and electric cables in the "old world"', but the shop floor view cannot have been so rosy.

On the site the only mechanised handling was on the wharf, where a crane running on train lines unloaded the barges. Everything else was man-handled. Four men would push a drum of lead-covered cable wherever it was needed. All the pigs of lead were pushed to the lead press on a flat trolley. Conditions were hardest in the drum shop, an old chemical factory where 14-year-old Jack Davis started at 4d (just over 1½p) an hour:

> The bare bricks were covered in a cotton-wool-like fungus. Dirt floor. Cloakrooms were unheard of. One's clothes were hung anywhere. This place was infested with rats and mice. The only way to protect your food and clothes was a tight wooden box with a close fitting lid . . . The work was very hard here, especially for me as I was no size and a lot of my work entailed carrying timber and hammering nails in. Some of these were four inch ones and it was no use tapping these so I soon had a sore shoulder and my right hand full of raw blisters . . . I was advised to bathe these in urine, which I did by holding my hand in the chamber when I went to bed. If I did not move it would soak all night. But gee, did it sting. Like a lot of things, it was crude but effective.

The heart of the works, where the Americans were conspicuous with their peg top trousers and their decisive 'yep', was more modern and impressed one of the post boys, Len Ballard:

> On the lower floor was the cable shop where the very noisy stranding machines were making telephone cable and power cable, massive great machines in sections with these big creelers holding the various wires and insulating tapes . . . Above the cable shop were manufactured cords for telephone sets and switchboards. The creelers whirled round at terrific speeds. Attended by girls. It was like a mill. How they stood the noise I don't know.

The foremen were martinets. Employee facilities were poor. Jack Davis again:

> A large dirty corrugated iron building, running alive with black beetles and crickets, was full of long wooden tables and forms to sit on. This was the canteen . . . Smoking and tea making were strictly prohibited in working hours. Everyone

> had a tea can. Mine was an empty baby food tin with a copper wire handle. An empty condensed milk tin, also with a wire handle, was the cup. A batch of tea made a pint. This was made in the morning and drunk cold the rest of the day. A batch of tea was tinned milk, tea and sugar all made into a sticky mixture and screwed into newspaper. Of course the paper got saturated so the mixture could not be scraped off. So the paper was put in as well and after a while was fished out with an iron wire spoon . . . Toilet facilities were again crude, as there was not the slightest privacy. The cubicles had no doors. The security were always popping in to see people were in the proper state of undress. If you were not, you were in for trouble . . . Outside the toilet was a long iron trough with four cold water taps. The trough always looked dirty and greasy and everyone was forbidden to use this during working hours.

The medical department was run by Tom Parry, who had a first aid certificate and wore a straw boater and a white coat. Nothing pleased him more than somebody saying 'Thank you, doctor', though his treatment was elementary. Young Jack Davis had his cuts and grazes attended to:

> The first thing he would do was scrub it with a small nail brush and disinfectant soap. This was not very pleasant. But while the scab was there this kept any dirt from entering the wound. There was a certain amount of common sense with his unusual ideas.

The employment and welfare department was more concerned with getting people than looking after them. Casual labour was hired for the day at the gate. As in the docks, likely workers were called out from those assembled, interviewed in a corner of the canteen and their hiring approved by a foreman. They would sign on with their mates – 'Bloody good cross this morning, Jim' – while those waiting at the gate were dismissed for the day. Working hours were from 7.00 a.m. to 5.30 p.m. with an hour for lunch, and from 7.00 a.m. to noon on Saturdays. To meet wartime demands there was also overtime and Sunday working.

When the men came home from the trenches, they wanted a better deal. After the labour unrest in the country during the war and the passions aroused by the Bolshevik takeover in Russia, the company was well aware of the mood. Even though many of the employees were female they would not necessarily be docile. Led by militants, they would want the world they had fought for. The company, which in 1918 came under the management of a new organisation, International Western Electric Company, sought further mutual support at home by joining the Engineering Employers' Federation. From 1919 it followed the Federation's policy of reducing the working week to forty-seven hours. It was a solitary concession. The week would not be reduced again until after the Second World War, but pay would more than once.

5

Merchants Have No Country: 1918–1925

Like other companies in the Engineering Employers' Federation, Western Electric was in a postwar trial of strength with the trade unions, culminating in the three-month lock-out of Amalgamated Engineering Union members in 1922. The employers won, showing that they were masters in their own houses and reducing wages to prove it. The wage of a skilled man in London came down from £4 13s 10d (£4. 69) to £3 0s 11d (£3. 04½). In the country at large trade was depressed and unemployment was rising but the fortunes of the company were running counter to the economy. The postwar period was one of growth and expansion.

After the Armistice, war contracts were soon liquidated and stocks of the relevant materials cleared. G. E. Pingree, who had been promoted from managing director of the British company to general manager of International Western Electric, was able to report on his old territory at the end of 1919: 'some difficulty in getting raw materials was being felt, but there was plenty of business and the output was improving from month to month'. He recognised that:

> The war brought home to many the importance of telephone service in the daily life of the people so that we find today a growing demand for efficient and comprehensive telephone service in all countries. In England our Associated Company is sadly in need of additional plant capacity. There is sufficient ground, however, at Woolwich, on which additional buildings can be erected, but Woolwich is becoming more and more an undesirable place for our manufactures, particularly telephone apparatus. The Woolwich neighborhood is changing into a big dock area and it becomes increasingly difficult year by year to find satisfactory labor for telephone apparatus manufacture. Our plans as regards the extension of

> the London Plant, however, are not quite formulated, but it is evident that in the very near future we shall have to add to our present plant capacity either at Woolwich or elsewhere.

Pingree's place as managing director had been taken by fellow countryman Henry Mark Pease, the former sales manager, a popular man who got on well with people inside and outside the company. He was not looked upon as an outsider by the Post Office, which was anxious to make up for lost time. The company was asked to undertake contracts on a 'work to completion' basis. This meant laying ducts, constructing manholes and installing the cable it had made so that everything was handed over ready for service. By March 1919 arrangements had been made for carrying out such work in eight of the fifteen superintending engineers' districts. For two or three years Foden steam wagons would be leaving Woolwich laden with cable for the Post Office.

An installation department was formed at Greenwich under E. S. Byng, formerly of the National and the Post Office and more recently involved in hydrophone installations. He set up a school to train jointers and testers for home and abroad, immediately for work on the Amsterdam–Rotterdam cable. New manual exchanges and extensions were supplied and installed in places like Liverpool, Bergen and Christiania. As partial assembly in the factory had not yet been introduced, much of the assembly and wiring of equipment was done on site by the installation department. Automatic telephony was beginning to come in. Rotary automatic exchanges made in Antwerp were installed at Dudley for the Post Office – an extension to the 1916 original – for the municipally owned system of Hull and for private users, as at Fort Dunlop.

Although the volume of business was welcome the shape of the future was uncertain. When ex-servicemen were selling matches on the street and employers were going bankrupt the company – it had joined the Federation of British Industries – had to secure its place in the postwar world, technically and with its major customer, the Post Office. In comparison with shattered Europe the UK was not badly off in its telephone system but, seen against the USA, it had stood still. The manufacturers and the Post Office had had to bend their energies to winning the war while, remotely, the Americans had proceeded with developments. Could their expertise be directly translated to Europe so that it could catch up?

Nobody saw the question more clearly than Frank Gill, who had just been appointed European chief engineer of International Western Electric, having been engineer-in-chief of the National from 1902 until its takeover. Like many of the pioneers in telecommunications, his formal technical knowledge had been acquired part-time. His real experience had been gained on the job, up poles and down manholes, and he wanted to give other young men opportunities to learn the technology of their time.

The component that was the basis of developments in telephony and radio was the valve, of which there were to be many types. As a future technical director put it: 'Every new valve is the most important discovery for years.' Ambrose Fleming, under whom Gill had studied, had in 1904 patented the thermionic valve, to which two years later Lee de Forest in the USA had added a third electrode, making the device into an amplifier. By amplifying signals that grew weaker and became distorted the further they travelled – much as pressure in a pipeline falls – it became possible to improve the quality of long-distance lines. Much of the necessary development work was done in the Western Electric laboratories in West Street, New York, out of which towards the end of the war came designs for repeaters that could be inserted at intervals in underground cables and overhead wires to amplify signals. Another technical advantage was that lighter lines would be needed. The introduction of the valve repeater was to reduce the copper content of long-distance cable from 300lb per mile to 20lb and to allow the number of conductors to be increased, with consequent benefits in production, installation and operation. The administrative consequences of the new development were important, as Gill realised: 'Those persons operating the various repeater stations must be operating to the same routine, employing the same technique and under the one control.' That statement held good, irrespective of political frontiers.

Valves were an important component in carrier circuits, which increased the capacity of long-distance overhead telephone and telegraph wires. Hitherto each communication had required one pair of wires. By using the technique of multiplex, these could now effectively act as a piggy back for a number of speech or data circuits, the separation between them being achieved by electronic filters. By the 1930s the technique would reduce the copper per circuit by a factor of more than ten. This was a relatively cheaper way of increasing capacity, especially on the long distances between US cities. Here again, a far-reaching advance in technology would know no frontiers. Chicago to Omaha or Pittsburgh was about the same distance as London to Hamburg or Geneva.

The immediate problem facing Gill was to transfer the American technology to Europe. In 1920 he decided to do it by sending his bright young men, his 'pups', to New York for nine months' training. Bertie Hinton, a fresh graduate, was one of at least seven who went over:

> It was equally necessary for the American Telephone and Telegraph Company to educate its own people as it was for Gill to get us educated. AT&T ran these courses for its own people and we joined them. There were about forty of them from all over the States who had come into the Long Lines Department of AT&T in Broadway. You spent two or three days with each expert: the man on filters, the one on coils, another on switching, ringing and so on. After this we went out into the field. Van Hasselt and Mack, for example, were specialising on carrier. I was on repeaters and I went off to Providence, Rhode Island, where a repeater station was being installed and actually worked on the installation. Then down to another repeater station at Princeton, New Jersey, installing and testing and getting the practical side of the business. When we came back we were not only equipped with books of theory. We also had a fair amount of practical experience. We actually made our own valves in West Street. Took the bits and pieces and put the whole thing together. I don't think it worked very well but it doesn't matter. We did it. We put together telephones, weighed out the carbon granules for the transmitter and took it from there.

When the young men sailed back to the UK they had tales to tell of evenings in Greenwich Village and information to impart to the senior research and development people in the Post Office. Bertie Hinton had all the confidence of his officer cadet and transatlantic training:

> They were all summoned together and a twenty one year old was set up to lecture them. I had to teach them. This information was not in Europe at all. We had gone and got it. We knew our stuff. It was no different from lecturing to a platoon.

Having no know-how of its own, the Post Office accepted Western Electric standards, equipment and prices. The first cable repeaters were installed at Fenny Stratford and Derby on the London–Manchester cable in 1923 and in the following year at the Wembley Exhibition the public was able to see on the company stand in the Palace of Engineering 'repeater panels with their glowing valves standing out from the sombre background'. The repeaters were switched on to an artificial line equivalent to a distance of 335 miles and strong and distinct speech was heard from telephone to telephone. When a repeater was cut nothing was heard. On this technical lead, the company was to build a major part of its activity, the transmission of electrical signals. For several years it was to be the major supplier of repeaters to the Post Office.

The company also introduced in 1924 iron dust loading coils that were about one half the potted size of the previous type. Improvements like this and lighter subscriber cable that could carry more conversations were welcomed by the Post Office because it was able to get more circuits into the available duct and manhole space to cope with the increasing telephone density in the business areas of large cities. But it was also aware that it relied too much on its suppliers for its engineering. It had no engineering department of its own to speak of and the only specifications and drawings it issued were of buildings. Repeater rack heights, for example, were usually to suit ceilings, with the result that bay frameworks varied from six to thirteen feet in height in three inch increments. Equipment layouts within a bay were therefore special for almost every contract.

On exchanges the Post Office was accumulating files of standard specifications, mainly based on the principles it had inherited from the prewar National. Most of the exchanges were manual, many of them obsolescent or inadequate. The few automatic ones were really experimental. Private automatic branch exchanges had also been introduced. There was no time like the present, with the telephone network beginning to grow again, to determine its future structure. The longer a decision was postponed the greater the problem would become. This was most evident in the capital and in 1919 two Siemens engineers presented a paper to the Institution of Electrical Engineers: *The Telephone Service of Large Cities, with Special Reference to London*. In the discussion that followed Sir William Slingo of the Post Office said:

> Although the capital cost is probably some 40 to 60 per cent higher in the case of the automatic than with the manual, I am

> satisfied that the annual cost of an automatic is less than the annual cost of the manual, mainly on account of the labour bill and the smaller premises required. But since operators' wages have risen so high, never to come back to their old level, there can be no question that the automatic system will justify its installation.

Everything pointed in that direction. The real question was which type of automatic system for local calls would be adopted.

When the Siemens scheme was not taken up by the Post Office two contending companies were left: Automatic Telephone Manufacturing and Western Electric. In terms of techniques for switching telephone calls this meant two different systems: the step-by-step developed from the Strowger patents and the rotary system. The Post Office had practical experience of both, but more of Strowger. Western Electric though had a new approach in the 'panel' system, which it had started to install in New York. McQuarrie, the company's chief engineer, came over to London and put its virtues to his opposite number at the Post Office, Colonel Purves, who later said:

> Mr McQuarrie's description of the new operating features which had been grafted on to the system came to me as a veritable flash of light.

Technically, it looked as though a coup were going to be made. Purves continues:

> Plans were prepared for a first panel exchange in London, to be known as 'Blackfriars', and at the same time the Post Office started to negotiate an agreement with the Western Electric Co., in accordance with which the manufacture of panel equipment would have been commenced in England under conditions that would ultimately have permitted other British telephone manufacturers to obtain a share of the work.

It looked as though the company, by far and away the dominant US manufacturer of telecommunications equipment, was going to secure the leading position in the UK. If it did, its equipment would be the standard for decades to come not only in Britain but also in the Empire, which followed Post Office practice. A major international prize of growing dimensions was within its grasp.

But the company made a fatal political error. It planned to manufacture most of the exchanges in Antwerp. How could Britain entrust the manufacture of nationally important equipment to another country, a country that it had recently gone to war to defend? The very factory in which it was proposed to make the exchanges had been occupied for all but a few weeks of the war by the Germans, who had stripped out most of the machinery and sent it to Germany. (In 1928 Western Electric was to be awarded $2,259,078 in reparations.) At the time the *Lusitania* was sunk and US entry into the war seemed likely, the Americans remaining in Antwerp had buried drawings, records and other items of value within the plant in zinc boxes, treasure to be recovered when the occupation was over. Memories of the war were still too fresh for a similar possibility to be entirely discounted.

Moreover, at home there was rising unemployment. Why put assured business for a public utility into the hands of a US manufacturer? The US had politically disengaged from Europe, refusing to become a member of the League of Nations, yet was prepared to take economic advantage. Unscathed by the war, it had been able to develop its economy and was now booming. Its wealthy inhabitants, conspicuous on their grand European visits, were resented. In particular, there may well have been some resentment within the Post Office that it was being led too much by a foreign manufacturer, that lacking the will, know-how and resources, it was having to listen too much to people like Gill's 'pups'.

Purves, who was responsible for recommending the type of system to be adopted, was a tough-minded man well aware of the crucial importance of his recommendation. He knew he would have the backing of the Postmaster-General and Parliament for a 'British' solution, even if this still involved American technology. In this he was helped by the sponsors of step-by-step automatic systems, who were pressing their case politically and technically:

> Early in 1922 the Automatic Telephone Manufacturing Co. called our attention to a notable development of the Strowger system by the Automatic Electric Co. of Chicago, who had succeeded in devising and combining with the step-by-step system a call-storing and translating scheme, which had endowed that system with practically the same elements of numbering and trunking flexibility that had first been conceived in association with the panel system. To this new

> development the name 'Director System' was given. The matter was, naturally, one of first-class interest and, although the immediate proposals were in a somewhat embryonic stage, they were at once investigated very fully.
>
> A working model was built and installed at the General Post Office and in November 1922 I definitely recommended the adoption of the step-by-step system, with the addition of the 'director'.

For this 'great degree of courage and resourcefulness' he was later publicly and unctuously congratulated by George Howard Nash, the chief engineer of the losing company. Among the reasons for Purves's decision were:

> The first cost of director exchanges was somewhat lower than the probable cost of the panel system manufactured in England . . . The fundamental electrical plan of the system is very much simpler . . . The Post Office engineering staff was already familiar with step-by-step systems . . . The Post Office was already committed to several variations of the step-by-step system on a considerable scale in the provinces . . .

His last reason was the most significant:

> Prior to the adoption of the director system, it had been ascertained that arrangements could be made for spreading the work among the regular exchange contractors of the Post Office at an early date. The important question of supply was thus greatly eased, as existing British factories became at once available for production purposes. The necessity for placing even the initial orders abroad was avoided, and the Post Office was able to enlist the co-operation of the skilled engineering staffs of all the contractors who had been producers of step-by-step equipment.

This did not mean that Western Electric was excluded – its parent had already secured US and overseas marketing rights on Strowger equipment. But its participation was on a much smaller scale. In countries where its rotary system was adopted it enjoyed half, perhaps as much as three quarters, of the total business. In the UK it had to be content with a monopoly in Hull. For the rest of the country it shared the business with Automatic Telephone, General Electric, and Siemens Brothers – all British companies. Patents were pooled so that a standardised system could be made and developments by one company could be manufactured by all the suppliers. The arrangement was formalised in 1923 with the first exchange agreement signed between the Post Office and the then four manufacturers. In 1927 they were to be joined by a fifth, Ericsson Telephones Ltd, with the result that each had a 20 per cent share of the business.

Manufacture of step-by-step equipment was based on the volume production of mechanical and electrical parts to fine tolerances. The mechanisms were much finer and more complex than those required for manual systems. They were subjected to accelerated life tests and called for new methods and equipment, for example higher quality press tools, the training of staff to adjust and inspect fine-limit relays, automatic test sets for checking circuit wiring. It amounted to the establishment of a new kind of home industry. As Purves put it:

> From the inception of the Post Office telephone system it has been the policy of the Department to discontinue the purchase of telephone plant from abroad and to encourage the establishment, within the United Kingdom, of adequate manufacturing resources for the supply of all its needs. It was well recognised from the first that this would foster the setting up of a British industry which would cater for a far more extensive market than that represented by Post Office requirements.

The odd man out was Western Electric. In trying to impress its system on the Post Office it had helped provoke an upsurge of nationalism. As a result, the British offshoot of the world's biggest telecommunications manufacturer had emerged as an integral but minor part of a new local industry competing in world markets. Those dealing with the company were aware that it was in some way responsible to its American parent. There were bound to be conflicts of interest. Internally that would raise problems of identity and loyalty and inevitably the problems of the growing child not being understood by its parent and, outside, doubts about who really made the decisions. Was the company British or an American stalking horse?

6

Going International – Gone: 1918–1925

The postwar period was a new era for another technology that had its origin in the late nineteenth century – radio. It was not taken too seriously by telecommunications traditionalists. 'Nobody did anything by wireless that could be done by wire' was the attitude. Radio had evolved from the experiments of Marconi and others to a means of telegraph communication, as with ships at sea. Its part in the 1910 arrest of the fleeing Dr Crippen was well known but as a technology it had not affected society at large. Speech had not really arrived. During the war Western Electric engineers built a large transmitter at Arlington, Virginia, for one-way radio telephone links but it was no more than an experiment. With a maximum of 300 of the 500 valves working successfully at one time, only a few words were understood at the receiving stations on the Eiffel Tower – a good radio mast – and in Honolulu, when conditions were favourable.

Conquest of these distances was a challenge, an opportunity for young engineers. The ether was free. Whatever they devised did not have to fit into some earthbound national Post Office pattern. It was steady, profitable work for customers like the Post Office though that paid for their experiments, as the cable and exchange men often reminded them. For the young, radio was a chance to pioneer, to learn from the parent company in the USA and perhaps do better. The bright young things could be as adventurous as they liked and society could catch up with their inventions. Their enthusiasm was concentrated on two main developments: transoceanic telephony and broadcasting.

Broadcasting in London began in 1922 at the eastern end of the Strand, where the offices of the electrical companies were concentrated. Marconi House was the site of 2LO and the

Western Electric experimental station 2WP was in the laboratory on the top floor of Oswaldestre House in Norfolk Street. The station, which had an output of about 75 watts, provided some authorised entertainment for the public and some unexpected interruptions for the staff. Transmissions were often picked up by the overhead telephone wires and a business conversation could suddenly include a ukulele duet that was being broadcast. At all costs the Post Office, which licensed transmissions, wanted to avoid any such confusion nationally. In particular, it argued that wireless operations had to be restricted so that there would be no interference with the communications of the air service and the navy. What it really feared was broadcasting following the lead of America, where stations were proliferating and air time was sold to advertisers.

As with telephone exchanges, a British solution was arrived at, a small group of official and protected contractors. Six electrical manufacturers together put up £60,000 of the authorised capital of the British Broadcasting Company, which was to derive its operating income from a half share of the ten shilling (50p) wireless licence and a ten per cent royalty on the selling price of receivers – imports were barred for two years. For its station, Western Electric ordered from the USA a 500-watt transmitter, which was temporarily installed at Oswaldestre House. The general office was converted into a studio by being carpeted and draped with heavy beige velvet curtains to stop echoes and an upright piano was hired. One of the secretaries, Miss Bott, made the announcements and records were broadcast by placing the microphone in front of a gramophone. The microphone was an improved type designed by the company, with a better frequency response than the usual telephone-type microphones.

Although it was unregistered, the BBC started its service on 14 November 1922. On Saturday 11 November the 2WP equipment was uprooted from Oswaldestre House and taken by steam lorry to the GEC works at Witton, Birmingham, where because of fog it arrived the following afternoon. No time had to be lost in getting the first provincial station going. It was installed while decorators distempered the walls and ceilings, the roof aerial was erected in fog, and the motor-generator set rested on a pile of doormats. After an hour's preliminary testing, broadcasting from the 300-square-foot studio began at 5.00 p.m. on Wednesday, 15 November, 24 hours behind 2LO. A. G. L. Mason, who a few months before had graduated from Birmingham University, spent the evening cooling the overheated motor-generator set with a grease gun. It was all just in time for broadcasting the general election results. A few days later the call sign was changed to 5IT. Ten days after the opening a Children's Corner was introduced and the engineer in charge, A. E. Thompson, who made the opening announcement, got a great kick out of becoming Uncle Tom. He claimed to be the first 'Wireless Uncle' but admitted that in the spontaneous programmes the engineer-broadcasters probably enjoyed themselves more than did the listening children. With the other engineers who had set up the station he soon returned to a Western Electric career.

On the engineering side the company helped to extend the range of BBC programmes. Because it had most of the expertise for ensuring good quality transmission over the cable circuits between sites and studios, it made the technical arrangements for the first outside broadcast, of Mozart's *The Magic Flute* from the Royal Opera House on 8 January 1923. Years later Peter Eckersley, who had started broadcasting for Marconi at Writtle in Essex and became the first chief engineer of the BBC, wrote:

> Indifferent, and still in a spirit of mild and amused tolerance, I put on the 'phones and tuned to London at 8 o'clock or thereabouts. Directly I put on the 'phones my whole attitude to broadcasting changed, and I have never forgotten the thrill with which I suddenly sensed the feeling of a large auditorium and was translated from the prosaic interior of the Writtle hut into the front row of the stalls at Covent Garden. When the music itself came on I sat absolutely amazed for three-quarters of an hour, and from that day to this my belief that broadcasting has a great artistic future has never wavered.

Encouraged by this success, the BBC wondered whether it would be possible to link its isolated stations for a simultaneous broadcast. Two months after the first broadcast from Covent Garden it asked the Post Office and Western Electric to carry out a test between London and Birmingham to see whether music could be transmitted over greater distances on the ordinary telephone circuits. The results were better than expected, the only serious distortion being through the loss of the higher harmonics of the violin and the top notes of the piano. Following another test between

Glasgow and London, a full-scale simultaneous broadcast was staged involving all six stations: London, Birmingham, Manchester, Newcastle, Cardiff and Glasgow. For the BBC it was a success, but not for the Post Office. To provide a strong signal from the broadcast studio, a Western Electric 40-watt public address amplifier was used to feed the outgoing lines. Its tremendous output caused a lot of cross-talk on the trunk lines and disrupted telephone traffic. On the next test six of the company's loudspeaker amplifiers were used, one for each outgoing line.

From the end of August 1923 simultaneous broadcasts of news, music and talks became a regular feature. The most notable was the opening of the British Empire Exhibition at Wembley on 23 April 1924 by King George V, the first broadcast by the monarch. His speech was heard by the throng in the stadium over a Western Electric public address system and relayed by the BBC. Estimates of the number of people who heard the King's voice for the first time ranged between five and ten million.

The company's public address system had been introduced to the British public the previous year with the words 'Hello! Hello! Buckingham Palace speaking' when HRH the Duke of York opened the Marine and Small Craft Exhibition at the Agricultural Hall, Islington, from the Palace. Every word he spoke was distinctly heard and in return he heard the applause of the crowd. The biggest problem was convincing his comptroller and equerry that it was unnecessary to shout or speak close to the transmitter. In Sheffield HRH the Prince of Wales addressed the people over it as 'Comrades'. Public address was used for transmitting speech and music at exhibitions, sports meetings (Wimbledon from 1926), election gatherings, religious services, railway stations, banquets and carnivals. At Southampton tests were made on directing ships into dock. Used for promoting products, it was known as 'audible advertising'. As the idea became established permanent installations were made in various public buildings like the Albert Hall. The Kone loudspeakers for them, made to the US pattern and an advance on horn loudspeakers, were continuously tested on a roof at Woolwich with two records: Schubert's *Unfinished Symphony* and Greig's *Solveig's Song*. Like John Reith, the first general manager of the BBC, the company was encouraging a popular interest in classical music, though it was scarcely music while you work.

From the beginning the company, in its own estimation, had seen that broadcasting would

> . . . add to the amenities of social life. With commendable promptitude wireless head receivers and simple inexpensive crystal sets were produced; over 7,000 being on the market before the official broadcasting licences were available. The excellent qualities of both the head receivers and the crystal sets were quickly appreciated by the trade and the influx of orders became so great that a scheduling system had to be adopted whereby a certain quantity of goods were despatched to the customers in strict accordance with the size and date of the order.

The retail price of the sets, popularly known as 'cat's whiskers', was £4 12s 6d (£4. 62½), and if they created a funny noise that could be heard it was enough to make the sale. Volume was catered for in the next development, valve sets using a loud speaker, sometimes housed in reproduction Chippendale:

> These cabinets were almost perfect pieces of craftsmanship and may be classed not merely as wireless sets, but as artistic furniture which will enhance the appearance of the most luxuriously appointed reception room.

Quality came in with the superheterodyne receiver, introduced by the company in 1923. For the humbler home there was the 'Weconomy' Wireless Set, a two-tier box with a large horn on top. Its ugliness was not diminished by the chief engineer's idea of having a bunch of flowers painted on it.

The venture into consumer products meant increased problems of security. For the first time the company was manufacturing items its employees could value. When a picture was taken of the workers massed behind the open gates at Woolwich for a 1923 publication to mark *The Fortieth Milestone*, the photographer asked them to close up so that he could get everybody in. Appropriate to the occasion, forty wireless sets were smuggled out. Even more vulnerable were Wecovalves, the two-inch-long receiving valves that used less current and were dubbed 'peanuts'. Gatekeepers used to search employees suspected of taking them out but as in most fiddles more ingenuity was applied than in doing the normal job. One method was the cut-out book. Having made his set, a thief could then get extra kudos by boasting to his mates about the faraway places he had been able to receive. 'Bloody liar' was a frequently murmured reply.

Not that the company was entirely honest in its valve sales. On one occasion it ordered 10,000 American-made valves from Montreal to take advantage of the lower duty on Empire goods. Engineering inspection's verdict was: 'We've been sold a bunch of stumers'. They were not uniform, some being good amplifiers, others performing well on the detector circuit. The stores manager, Reg Ballard, who had an eye for a sale and some traffic lights in a stockroom because the company was thinking of going into that business, came up with a solution: sorting the valves, marking them with a red, yellow or green spot of lacquer according to their function, and pasting an equivalent spot on the carton. With a bit of publicity, it worked. Soon after, the company got its own production line going successfully. Although the valves could be used on any circuit, customers insisted on ordering by colour. For at least two years a small group of girls was employed putting on spots to suit the orders coming in.

Establishing a valve department involved the introduction of new equipment and processes for glass working, heat treatment of metal parts, electric spot welding, the coating of platinum alloy wire to make filaments, and vacuum pumping. A particularly high vacuum was required for the large transmission valves, the manufacture of which followed a year after those for radios and repeaters. Using much more power, the transmitting valves were water cooled. Valves of all types were good repeat business, the argument being that what the company really sold were sockets that had to be kept filled with working components.

At a time of general deflation new technologies and new and enlarged markets added up to more business and the need for more space. The growth was genuine, unlike the illusion of much postwar prosperity. Even so, the company took a prudent course, at first taking on temporary premises and expanding on its existing site before acquiring new plants and offices. In 1920 a lease was taken on a small ex-Vickers plant at Bexleyheath, where there were shops for iron and wood work, wiring and assembly of switchboards. Later, radio cabinets were made there. As at the Woolwich site, where the lease of a glass bottle manufacturer expired in 1921 and a 16,000-square-feet building was taken over, some of the working conditions were primitive. F. W. Jarrett, who joined as iron shop storeman directly from school, had little training and learned by trial and error:

> I was given a table in the blacksmith's shop and a store cupboard. Every surface had a permanent layer of soot. Even there the smell of burning oil we used in finishing screws, nuts and washers could not be endured. We had to do the job in the yard. The old blacksmith, Charlie Andrews, looked after me, and my lunch sandwiches were toasted by being held under a red hot bar of iron. We could buy from the canteen slices of bread and dripping for ½d (less than ¼p).

Conditions were initially not much better in the new headquarters leased at Connaught House, Aldwych, which the company began occupying from November 1922. Here were housed the sales and engineering departments of the British company and the European headquarters of International Western Electric. Staff moved in while building was still going on and had to get the plaster dust off themselves every night.

Sales and the number of employees were rising but not *pari passu*: sales per employee were falling. More people had to be taken on to produce equipment for the expanding Post Office and overseas markets. As a successful member of the engineering industry paying the lowered wage rate from mid-1922 Western Electric could afford to employ on a generous scale. At a time when interest rates were high and credit expensive it also had cash to take advantage of the difficulties of other engineering companies. The one it lighted on was J. Tylor & Sons Ltd, which during the war had started to develop a site at rural New Southgate in North London for the manufacture of lorry engines. By 1921 it was in liquidation.

For £80,000, in 1922 Western Electric acquired from the receiver a 27-acre site with a two-storey concrete building by the Great Northern Railway line out of King's Cross. It was an even better bargain than the purchase of the Fowler-Waring cable plant in 1897 for £87,000. There was plenty of opportunity for modernisation – men and women washed in the same troughs – and room for expansion on the site. The main product there was to be telephone equipment for the Post Office, which had come down firmly against buying exchanges from Antwerp. Indeed, the first people to move into the new quarters were from Antwerp, the patent department of International Western. One of the perceived advantages of the rural location – it was not far from the estate of Sir Thomas Lipton, grocer and yachtsman – was that newly recruited labour would be less union-minded than the employees in dockland.

Large as the Southgate site was, the building was not enough for immediate needs. In 1925 and 1926 short leases were taken on properties at Hendon owned by an unsuccessful aircraft manufacturer, the Grahame-White Company, and by London Aerodrome. Total floor space of the single-storey buildings was almost 400,000 square feet. Manufacture was transferred there from Bexleyheath, leaving the company with three major plants in North and East London. It was recognised that this was not the most efficient form of organisation and the intention was to concentrate all activites in one place. For this purpose in 1925 the company started buying parcels of land at Wembley.

It was raising its sights to broader horizons, largely under the influence of Frank Gill, the European chief engineer, who was responsible for the planning and carrying out of long-distance communications projects. Like Kingsbury, one of his closest friends, he was a missionary for the telephone. As an international consulting engineer he had made acquaintances in many places and gained a reputation as an upright Christian – he was a churchwarden in Beckenham – who returned the gifts of grateful clients. Conducting his business on strict principles, when he was on his way to a directors' meeting of the Constantinople Telephone Company at which the dividend would be raised he would not answer his mother's question whether she should increase her shareholding. If he could not impress his principles throughout the organisation – the prewar international business-sharing agreement between Western Electric and Siemens & Halske was renegotiated in 1921 with the local and not the parent company acting as the contracting party – he could at least lift eyes to new business opportunities.

Branch offices were opened in Simla, Singapore and Cairo but it was in Europe that most of the activity took place. Here Gill was able to apply the technological advances that had been made in the USA during the war and imported via his 'pups'. The first long-distance telephone cables using repeaters were planned, made and laid between Gothenburg and Stockholm, on the Milan–Turin–Genoa route, and between Paris and Strasbourg. Norman Kipping, who was to be important in the company's interwar development, as a junior engineer made the first telephone call from Stockholm to London. He proposed to his fiancée Eileen, who was too dumbfounded to reply. A tentative step was made outside Europe into Africa with a circuit between Madrid and Tetuan in Morocco, as a strategic weapon in the French and Spanish fight against the Riff rebels.

Significant as these achievements were, Gill realised that they had to be part of a grander scheme. In the year of Bell's death, 1922, during his inaugural address as president of the Institution of Electrical Engineers, Gill asked two questions:

> Has telephony, during the 46 years it has been available, been of as much use to Europe as it might have been?
> Have the organisations, Government and otherwise, been permitted to do what they wished to do? The answer to both questions is most decidedly – No!

In particular, the through business was 'meagre in quantity, slow and inefficient', which contrasted ill with the Bell system, a view he emphasised with a map of Europe on which he superimposed American distances:

> The direct distance between Brussels and Athens, or between Paris and Constantinople, is 1,300 miles – about the same distance as between New York and Omaha, or between Chicago and Salt Lake City, between which places calls can be made at any time. The direct distance over land between London and Baghdad is about the same as between New York and San Francisco, over which line conversations take place daily.

In comparison, the UK was connected to the Continent by only twenty-three circuits, many of doubtful efficiency.

So that the through business could be handled properly it was necessary 'to depute a body to do for all European nations that which no one nation can do for itself'. He concluded by suggesting that the telephone authorities of Europe should hold an early conference to study the problem and endeavour to find a solution. His eloquent plea to a UK audience inspired concerted international action. The following year at a technical committee meeting in Paris it was unanimously agreed that an international consulting committee should be established. This came into being in 1924 under the title Comité Consultatif Internationale des Communications Téléphoniques à Grande Distance, fortunately abbreviated to CCI. Many found it easier to refer to it as the Gill Plan. In effect a communications League of Nations, and like it having no executive authority, the CCI was to be more enduring and successful.

The formation was none too soon, for the telephone could now span oceans. On the night of 14-15 January 1923 the first transatlantic radio–telephone call was made. Beginning at 9.00 p.m. New York time, 2.00 a.m. GMT, it was a two-hour one-way event. Thayer, the president of AT&T, and others talked to Gill and a group from the company, the Post Office, and the press assembled in a wooden hut on the Southgate site, where dress was dinner jackets and earphones. Individual voices were recognised and successful reception reported back by cablegram. Towards the end, Gill sent a message to Thayer: 'Loud speaker now being used – good results – great enthusiasm. Your interview on loud speaker came through fine.' In the enthusiasm nobody had arranged to record the event, but J. E. Kingsbury, who had left the company because of a disagreement with Thayer and had returned as a director, was there and gave them his shorthand notes.

Among the guests was Marconi, which gave many the impression that he had been responsible. In fact the transmitter, at Rocky Point on Long Island, was the property of the Radio Corporation of America and the demonstration followed weeks of tests to check the quality of reception under different conditions. At the conclusion of the experiment Thayer was careful to point out 'there is much more to be done before we can speak definitely about establishing practical commercial radio–telephone service across the Atlantic'. For more than a year a team of fresh graduate engineers investigated the behaviour of radio signals across the ocean in terms of wavelength, time of day, season of the year, intensity of atmospheric disturbances, and the like. One of the engineers was Maurice Deloraine, who had served part of his time in the French Army at the Eiffel Tower on the wartime transatlantic experiments. The group he was now with did establish, they believed for the first time, a correlation between the quality of radio signals and the intensity of sunspot activity.

Clearly a new age of international communications was coming. Less discernible were the attitudes and organisation that would accompany it. As the world began to shrink to a global village it would need leaders of broader outlook, citizens of more than one country. Successful though it was, Western Electric was not that kind of company. It was well organised. The departmental and part numbering systems, the company orders (e.g. 'What to do when the contract comes in') were to survive for decades. Prewar it had the tight discipline of monthly and quarterly reports to be in London by specified dates. The reports were not only of value to the parent company but also intended to help local managements run their business. In Berlin, Budapest, St Petersburg or wherever 'the Allied Interest' was, they could look at their net sales, number of employees, orders, receivables, distribution costs, shop statistics, trial balance and so on and know where they stood. A secret code book existed for preserving the confidentiality of communications on commercial transactions. At the same time the organisation was mindful of its employees' welfare. As long ago as 1906 it had instituted a pension scheme depending upon service not status. Being a Westernite gave one some status in a community.

It was a solid, conservative company, not what many people thought of as typically American. It had about it though an inflexibility, a sameness. Its plants in Europe were a scaled-down version of the Hawthorne works in Chicago. North America was where the money was really made. Keeping an eye on overseas business required a disproportionate amount of management effort. Executives enjoyed the travelling – it was much more civilised getting a drink on a Cunarder or a boulevard than from an Al Capone speakeasy – but it was not necessarily the best economics. Postwar, the isolationist mood was strong too, not just in the Middle West.

There were, however, in the USA two brothers, cosmopolitan in background, who had a different outlook. Hernand Behn and his younger, more dynamic brother Sosthenes had been born of French–German parents in the Danish West Indies, educated in France, and in business as New York bankers and sugar factors in the Caribbean. There they acquired telephone companies in Puerto Rico and Cuba, entities with a total of some 40,000 telephones that in 1920 they formed into International Telephone and Telegraph Corporation. Unlike Edison, Westinghouse and other nineteenth-century entrepreneurs, they did not name the enterprise after themselves. Their concept, on a much flimsier base, was altogether grander. Perhaps the name was intended to convey the idea of a big brother to American Telephone and Telegraph, with which Sosthenes had concluded an agreement for the laying of a jointly and equally owned undersea cable between Havana and Key West. The gentlemanly adventurers of the Spanish Main were now ready to launch out on an international scale and conquer Spain itself.

Having experienced the value of the telephone in the Moroccan campaign and realising that good communications would help bring unity to their own divided country, King Alfonso XIII and the dictator Primo de Rivera let it be known that they would consider offers to develop and operate the national telephone system. In princely style Sosthenes Behn took a suite at the Ritz in Madrid, from which he organised a band of engineers, accountants and typists to prepare proposals. What he was selling was confidence. He had no factories, a small technical staff, and his wealth did not equal the impression created by his entertaining – the reputed box of gold he carried around contained his favourite cheese – but he impressed the top people who mattered right up to the King, whose signature in 1924 gave him the concession.

The snag was that a substantial amount of the manufacture had to be in Spain which, like Britain, wanted a national industry. Behn's defeated competitors, Siemens and Ericssons, were not anxious to oblige and Western Electric was not interested in adding to its less profitable European interests. To save his bold deal and guided by prestige rather than profit, Behn drew an enormous bow at a venture. He offered to buy International Western Electric. His proposition played into the hands of the new president of AT&T, Walter S. Gifford, a laconic Yankee accountant who was threatened with government action over his US subscribers subsidising his foreign interests and who wanted to concentrate on developing AT&T as the best domestic operator. Gifford asked a high price, $30 million (£6.1m), for a group of dispersed companies that produced a fluctuating profit and loss record and had no manufacturing in Spain. It was the same price the USA had paid Denmark in 1917 for the Virgin Islands, the Behns' birthplace. Behn raised his money from J. P. Morgan and Company. Out of the $29,306,534 (£6,050,381) cash received, Western Electric made a net profit of $15,955,096 (£3,293,966).

For his outlay, Behn gained on 30 September 1925 manufacturing plants, technology and people. In the agreement on the continuing exchange of patents and technical manufacturing information in the arts of telephony and telegraphy ITT had acquired much needed expertise for use outside North America. The two main exclusions were sound equipment for 'talking motion pictures', which were to become commercially practicable the following year, and undersea cables, in which Behn had a lifelong personal interest. People who had come under International Western Electric tended to stay with the new parent, International Standard Electric (ISEC). For example, Frank Gill remained as the European chief engineer and H. M. Pease as the managing director of the British company.

One director who resigned was J. E. Kingsbury, who had founded the local organisation forty-two years earlier. Six days before the takeover he wrote to the company secretary:

> The International Telephone and Telegraph Company who now control the Western Electric Co Ltd have interests and intentions in the operating field which did not exist under the previous ownership. In the carrying out of their programme it is possible, even if not highly probable, that there may be some overlapping of what I may perhaps conveniently call political issues with other interests in which I take an active part. In such case the holding of an official position on both sides is not consistent with the proper carrying out of one's functions.

The following day his resignation was accepted with deep regret and the confirming letter from the managing director concluded: 'I am sending you the enclosed cheque for £3,000 which I trust you will accept as an appreciation of the Management for services rendered in the past for which you have not been adequately remunerated.' At the age of 70 Kingsbury, an authority on telephony, went on to continue his career in another new technology, plastics. Behind him he left a company that was to change its character considerably.

7

Under New Ownership: 1925–1930

The name of the British company was changed to Standard Telephones and Cables Limited but its continuity was emphasised by using the Western Electric typeface, the 'flash' characters. To the world at large it appeared not a new company but a relative of the one that had long been advertised on the sides of buses and cable drums and that would soon appear in the credits for talking pictures. It was a change of form rather than substance. The organisation that had been created carried on with its company orders and numbered departments in much the same way. Indeed, fifty years later when the stores at Southgate were reorganised some Western Electric part numbers were found to be still in use. In 1925, after minor apprehensions about their future, the employees noticed no essential change in their day-to-day lives. Their conditions of service remained the same but some felt that the name Standard, although intended to imply excellence and reliability, lowered their status somewhat and put the organisation on a par with similarly named companies. It was not quite the organisation they had joined and also an inaccurate description. The company would soon cease to make complete telephones and there was no mention of its burgeoning radio work.

A change in the status and character of the organisation was, however, to become evident. Whereas the former company had been a comparatively small manufacturing outpost of its American parent the new company was of some substance in the grandiosely named International Telephone and Telegraph Corporation, which was really a suite of offices and a dominant personality. In line with Behn's strategy of building national companies with a prime local loyalty – he reserved the role of international *grand seigneur* to himself –

STC became more British. It still had access to Western Electric technology, and initially largely relied on it, but it also had the opportunity and need to develop on its own. The adopted child, out of its sheltered environment, could grow to meet some of the demands of Colonel Behn, whose only American resources were financial.

Actually the Colonel was his own minor promotion; he had been a Lieutenant-Colonel in the Signal Corps in the First World War. A tall man, he retained an impressive military bearing. When he sailed from New York his departure was noted in the quality London papers and his local staff got everything shipshape for the arrival of the president. He liked staying at the Savoy, where his own soap was provided. Doormen at the hotel had an arrangement with their opposite numbers at Connaught House to let them know when his car had left and, within two or three minutes, he would set foot in the Aldwych offices. In readiness one of the two elevators – he appreciated the American usage – was kept on the ground floor to take him up to his walnut suite, otherwise used only for board meetings. Normally he would arrive in the late morning, have a brief meeting with the managing director, go off to the Ritz for a lunch that was preceded by a daiquiri and included his selection from the cellars ordered from the sommelier, whom he knew personally, and then back to the Savoy for an afternoon nap to leave him fresh for work in the evening. His work was that of the diplomat and entertainer winning friends and influencing people. As one employee privately observed: 'Of course ITT is not really a commercial company at all. ITT is a principality. It has a prince, it has a court.'

His management style was personal. A great traveller, Behn was easily at home anywhere, visiting offices or plants, seeing what was going on and feeling the atmosphere for himself. His memory retained key figures and decisions. Charts purporting to show trends he could not abide, claiming that he could not understand them. What he wanted were the figures plainly listed. Not that he was a man to pore over figures. He was much more a person of ideas, visions to be realised, the grander the better. Unfortunately sometimes the laws of physics got in the way and the engineering was painfully slow.

The short-lived 1926 general strike was not a setback to his ambitions. Staff and non-union employees carried on working. Some helped the London and North Eastern Railway in keeping its communications going, help that they believed was later rewarded with orders for train despatching equipment. For the company, business was still growing and it was doing its bit to make it grow even faster. With other members of the industry it had in 1924 formed the Telephone Development Association to promote the virtues of the telephone directly to the public. When characters in the *Strand Magazine* were still sending wires, especially in emergencies, an earnest group at one end of the street was trying to change public attitudes. A piece of enlightened self-interest, it was an early example of a successful public relations campaign.

The Association produced advertisements, posters, pamphlets and a cartoon film, gained press and radio publicity – popular songs helped – and held meetings to put over the idea that the telephone was not a luxury but a matter-of-course necessity. One of its advertisements pointed out:

> To have a telephone in the house costs you
> NOTHING to install
> NOTHING for the calls you receive
> A PENNY for the calls you make
> 2/1½d [under 11p] A WEEK for rental

A pamphlet began:

> THE FRIENDLINESS OF A 'PHONE
>
> It's a wonderful pal. Are you lonely? A ring will put you in touch with friends. Nothing doing tonight! Prr—ing! An invitation to the theatre or 'Come on over and make up a four at bridge.' Daddy lost the last train! Well, any way, it's good to hear his voice – to know he is all right – to have a car at the junction.

and included little drawings with captions like 'Gracious! Jack's bringing a friend to dinner. I must 'phone the stores.'

Besides being good public relations it was sound Keynesian economics. In 1925 the Association argued that 'every million pounds per annum expended upon telephones provided employment for, approximately, 6500 people, including trades directly and indirectly concerned'. Arguments like this secured the support of trade associations, companies, and the Federation of British Industries. Of course the manufacturers, who were doing the promotion the reticent Post Office ought to have been doing for itself, stood to benefit. Every telephone installed increased demand for cables and apparatus, which in turn meant business for the metal, textile, paint and other trades. Almost virgin territory for telephone development was the English village. Down at Montacute House in Somerset in 1923 Lord Curzon heard by letter that he had been passed over as Prime Minister. Montacute was not on the telephone.

In 1925 the company completed the backbone of the UK trunk network by installing cable and repeaters on the London–Glasgow route. Four years later it introduced loading coils with cores made of permalloy, which gave them better electrical properties and enabled the volume of the case to be halved again. Between 1925 and 1930 it installed automatic exchanges in Bedford, BERmondsey (a London subscriber's number consisted of the first three characters of the exchange followed by four digits), Bridlington, FULham, Halifax, Macclesfield, Maidstone, Manchester, Oxford, Rochdale, SLOane (a 10,000-line exchange involving the wiring and soldering of over a million connections), Southend, and TEMple Bar. Often a number of exchanges were installed in one area, with the installation staff moving from one site to the next. A large manual installation in 1927 was the second toll exchange opened to relieve London trunks of traffic within a 50-mile radius and thus speed up long-distance calls. With positions for 202 operators and working in conjunction with the first London toll exchange the company installed in 1921, it handled over 60,000 calls a day.

Rotary exchanges continued to be manufactured and exported. In 1928 contracts were secured for about 30,000 lines in Cairo and Alexandria, for which the company undertook to train personnel of Egyptian State Telegraphs and Telephones in installation and maintenance, which also involved unofficial lunchtime boxing practice. The perk of the minister responsible was his then customary half per cent of the total contract value. In Dublin there were problems with the new type of wax used in the relays. It melted and the exchange seized up. A team of young engineers was equipped with toothbrushes and buckets of solvent that dissolved the toothbrushes more quickly than the wax. A more fascinating posting from all points of view was for the installation in Shanghai.

The big international developments though were in radio. After the successful demonstration of one-way transatlantic radio-telephony in 1923 a series of measurements had established the feasibility of a commercial two-way service almost round the clock. To match the radio station at Rocky Point the

Post Office decided to build one at Rugby, for which the company was awarded the contract to design and supply the radio telephone transmitter. Its construction was an international effort. The team was led by a Frenchman, Maurice Deloraine, who spent five months in the USA, and a Dublin graduate, Charles Strong, working with some Western Electric engineers. The average age of the group was about 25. The 200-kilowatt longwave transmitter was installed in 1926 and, after considerable difficulties with the high voltages successively blowing about twenty-five £100 valves, went into commercial service on 7 January 1927. A three-minute call cost £15 and initially there was only one circuit serving the London and New York metropolitan areas but, with the radio-telegraph service started on 1 January, it was enough to make it transatlantic year. Another 25-year-old, Charles Lindbergh, made his solo crossing by air in May.

Francis McLean, a future BBC director of engineering, had joined the company in 1925 and seen it change over just a short time:

> It was a very creative period. It was also creative in that the European-based people, particularly people in London, were starting to do things on their own. When I first joined the company it was largely a question of taking the Bell Labs' spec and changing the dimensions just a bit, the power supply or something, and building it and hoping it would work. But we started doing independent work. We made a five-kilowatt broadcasting station, which we put on display at the big exhibition held in Como in 1927 to mark the centenary of the death of Volta. We felt we were doing something. We no longer felt we were copyists. We were originating things.

When the Atlantic had been spanned by long wave it soon became evident that more than one circuit would be needed. Short wave seemed the answer. It was not as reliable as long wave but it was cheaper and a combination of the two systems could provide a transatlantic service over the full twenty-four hours of the day. Making short-wave transmitters though was something entirely new and there were many problems to be solved. Charles Strong was sent over to the USA to investigate:

> Although Western Electric let us see what they were doing, they themselves weren't all that far advanced. They weren't all that sure about their transmitters. So we in fact were developing them at about the same time and not very much behind.

With the enthusiasm of the young, work proceeded quickly. If the young engineers stopped to reflect they could even feel at times that they were doing something for humanity, bringing people closer together. Their work impressed the chief engineer of the Post Office enough for him to recommend that STC be awarded the contract for the first short-wave transmitter for Rugby. This provided an additional channel to the USA from August 1928 and the basic rate to New York was reduced to £9 for a three-minute conversation. By 1930 there were four circuits in operation, three short-wave and one long-wave, with another long-wave transmitter under construction.

The significance of these developments was appreciated early by Behn, anxious to give meaning to his newly enlarged international corporation. He had technical teams established in Paris and Hendon to work simultaneously. Paris was to be responsible for overall system engineering while Hendon designed and made the transmitters and receivers. Their work gave further occasion for Behn to impress the Spaniards that they had made the right decision in coming to ITT. In 1929 the world's second transatlantic link went into service, between Madrid and Buenos Aires, with an onward short-wave link to New York following later.

Broadcasting too offered an expanding international market for transmitters. The 5-kilowatt medium wave transmitter built at Hendon on speculation as a saleable development model was eventually installed at Wellington, New Zealand, in 1927. By 1930 some thirty broadcasters of powers up to 15 kilowatts were built and installed in Europe, South America, Australasia, Japan and South Africa. The number of long-distance telephone circuits was increasing too with the application of carrier techniques. At this stage three extra circuits could be piggy-backed on each pair of overhead wires. In 1925 the first three-channel carrier was manufactured and installed between Cape Town and Durban. The installation was done by the South African administration under the supervision of an engineer from England, Leslie Long:

> Van Hasselt and Montgomery had been to Western Electric in America before the takeover, Van to study the transmission line problems and Monty the apparatus side of the thing. When they got back they started copying the American three-channel system. I was the poor boob sent out to South Africa to install it. In effect it was a field trial, even though it was nearly 6,000 miles away. It was an act of faith. Apart from a

> component problem – the paper-wound condensers kept breaking down – we didn't have any difficulty with it. From there I moved on to Spain, planning the use of the system there. We had an international lead in carrier systems and I got involved in projects all over the world.

The Australian Posts and Telegraphs Department had a system installed on the Melbourne–Sydney route, where there were only two congested telephone circuits in use and there could be delays of over an hour in getting through. It was so successful in improving the service that by 1930 about fifty systems were operating throughout the country on some 10,000 miles of trunk lines. Spain had eighteen systems and there was a route from Madrid to Paris. Other territories to have three-channel systems installed included the Argentine, Brazil, Federated Malay States, Japan, New Zealand, Norway and Poland. One experimental system was sold to the Post Office for field trials across the Highlands between Inverness and Ullapool but there were no more home installations.

Clearly the success and growth of the company depended upon technical innovation. Although it had access to much Western Electric technology it needed to establish and develop its own resources. Technology was advancing and the company could not afford to fall behind. Conscious of his prestige, Behn also wanted to show that he could be independently successful. Hence he established two laboratories, the British one in Hendon and the French one in Paris. Broadly, Hendon would concentrate on advances in cable and Paris on exchanges and radio, but they would work together. The teams were not exclusively British and French and Behn was not bothered about what they did in detail as long as they achieved spectacular results quickly. Until they did he could boast that outside the USA he had the largest organisation in the world devoted exclusively to research and development in electrical communication.

At Hendon the laboratories were in the premises of the former London Country Club, across the road from the manufacturing plant. The ballroom became the laboratory and the bedrooms off the first floor balcony executive offices. A good time continued to be had by the new occupiers, grey-flannelled young men who enjoyed taking their Morgan cars at speed round the oval flower bed at the front or playing catch with measuring instruments worth more than their £150 annual salary as a relief from the routine of performance tests on items like telephone transmitters and receivers. Even had they been more serious, they could not have achieved the technical coup that Behn wanted.

For more than a decade he had been fascinated by the possibilities of undersea telephone cables. In 1921 three had successfully joined Key West and his telephone company in Havana, a distance of ninety-nine nautical miles. Now that Western Electric – the UK company had been his guarantor for the supply of the Caribbean cables from Britain – had excluded the technology from his agreement, he wanted to beat them at their own game. His ambition in transatlantic year was no less than being the first across with an undersea telephone cable that would both rival radio and show that his new organisation was a match for the might of AT&T. When consulted on the project, Gill tried to discourage him. So Behn asked the opinion of Gill's assistant, Erikson, who was also against the idea. In the present state of technology it was not possible to achieve the right transmission characteristics for an undersea telephone cable over such a distance. It was a totally different proposition from the low-speed transatlantic telegraph cables that had begun in the nineteenth century or the short telephone distances in places like the Caribbean. Yet there was no point in going into the technicalities, trying to explain the four electrical coefficients: capacitance, inductance, leakance and resistance.

Not one to give up, Behn consulted Nash, who was chief engineer of STC, responsible for the labs, still trailed clouds of glory from the wartime 'fish' to which he had given his name, and had advised Behn on the purchase of his yacht *Aphra*. Short in stature, long on ambition, he said it could be done. Both men were at sea, wishing past successes into future achievements. Nash's confidential view could have come from Behn's lips: 'What you need now in this company is imagination and more imagination.' Mack, who was privately credited with the work on the 'fish' and on whom the technical responsibility for the cable would have fallen, was not consulted by Nash before the commitment was given. He regarded the project not as engineering but black magic. There were enormous problems to be overcome to make a cable that, if successful, would handle only one voice circuit one way. The direction of the circuit would have been switched by whichever voice was speaking. One problem was the weakening of the signal over the enormous distances between the northerly landfalls. No thought was given to undersea repeaters, with the result that the strength of the signal at the far end of the cable would have been about one ten-thousandth of what was desirable. Another problem was an insulating material with the right electrical properties.

When being recruited, Dr Malcolm Field had stuck out for £350 a year, which was a lot more than Nash paid his engineers, but he argued he was worth it because he was a chemist:

> The chemical laboratory was initially in the old kitchens. We were experimenting with rubber and rubber compounds with different types of waxes. Bell Labs were also working on similar lines and they actually got a few cables made with a type of rubber wax. We also tried to improve on conventional materials like gutta percha, which had been used for telegraph cables, and balata. On our list of hydrocarbon materials was the ideal one, something very stable like polythene. But we didn't have the equipment to make it. I read a paper that suggested it could be produced without very high pressures and I set the labs on fire trying to do it. We managed to put the fire out ourselves. We were on the track of the right material but we had no idea how to make it. It took the resources of ICI to produce polythene in its labs in 1933. Even then it was not available in sufficient quantity for experimental cable manufacture until 1938.

With the available technology, basically that for telegraph cables, the cable would have been several inches in diameter, very expensive, very heavy and very difficult to lay. It never got beyond samples soaking in brine in long tanks for life test. Other achievements were minor. Work was done on combining telephone transmitter and receiver in a handset to take the place of the 'candlestick' model but the Post Office preferred a competitive design. Improvements were made in the layout of equipment racks and development of a larger capacity system for transmitting telegraph signals was to pay off in the early 1930s. The less tangible but nevertheless real benefit of the laboratories was in strengthening the independence and confidence of the British engineers. Returns from that were not immediate.

In day-to-day business attempts were made to extend the range of products. Gamewell fire alarms were much more successful than the road traffic control system, of which there was only one installation, in Cairo. Medical items like audiometers and electrical stethoscopes were not a success.

The company also stopped making wireless sets. Nevertheless it continued to acquire space. Head office expanded with the lease of floors in nearby buildings in Aldwych. Further land was acquired at Wembley, bringing the total by 1930 to some 157 acres, and the site partly developed with sidings from the London Midland and Scottish Railway, a reinforced concrete wharf on the Grand Junction Canal, an artesian well, a reservoir, main services, a road and trees. E. S. Byng, who succeeded H. M. Pease as managing director in 1928, took a group of executives from the existing plants to the site: 'Gentlemen, this is your new home.' The intention was to consolidate the Hendon and New Southgate activities there, but Woolwich cable-making presented a problem because it was not possible to berth ships on the canal for loading long lengths of lead-covered undersea telephone cable.

Manufacturing as then dispersed was inefficient. Now that a 'ring' of five manufacturers, the Bulk Contracts Committee, annually shared out the growing Post Office exchange business they were into an age of mass production. In return for the guaranteed business though they were subject to cost investigation. Every five years the Post Office examined two reference plants, Automatic Telephone in Liverpool and STC, New Southgate. Shop costs and overheads were assessed, with arguments about whether canteen losses and the social and athletic club were a permissible charge, and unit prices negotiated on the range of standard parts that went to make up an exchange. These became the prices at which all five manufacturers had to supply. Hence there was an incentive to make parts more cheaply and keep one's manufacturing secrets from the other four. In this way it was possible to improve a percentage point or two on the agreed rate of return, which was about 7 or 8 per cent. The whole exercise was a battle of wits between the spenders of public money and industrialists trying to screw out extra multiple coppers.

In 1928, the year that the second five-year agreement was signed with the Post Office, Norman Kipping, the tall commanding works manager of New Southgate – he was nicknamed the Baron twenty years before he was knighted – spent four months in North America looking at production methods. Most of his time was spent with Western Electric in Chicago and he also visited Northern Electric in Montreal, Automatic Electric, the Stromberg Carlson radio receiver plant, the Detroit factories of Henry Ford, and Eastman

Kodak. In his eighty-page report, with detailed sections on organisation, machines and processes, he noted:

> The three most obvious and essential differences between the Western Electric Company and European conditions are, of course, the enormous outputs of the former, their high labour rates, and the fact that they are able to obtain fairly accurate advance information of the output requirements. The outputs are from 4 to 12 times greater than those of the larger European Houses. The rates are, on average, two to three times greater than in London for similar operations . . . Almost any unit of plant which may be studied has been brought into existence as a result of the combined efforts of many skilled specialists, with the sole object in view of saving or economy, or improved work. Frequently one finds automatic machines used for work which one had previously regarded as essentially hand work, but it becomes evident that the most detailed analytical methods have been used in studying methods of manufacture with the object of eliminating waste time, unnecessary work, and expensive labour.

In Western Electric he looked at unit costs to three places of decimal pence and came away convinced that costs back home could be reduced. He set a team to work on it and adopted many of the American techniques.

Within a few years of becoming STC, the company was beginning to establish its own identity. In 1928, for example, it started its own employee magazine, *Standard News*, and in the ensuing year instituted its own pension plan in preference to the parent company's. In developing, it was using some American money, it was learning from America but it was neither an American subsidiary conforming to type nor a lesser version of its parent. ITT was characterised by Siemens & Halske in 1930, which alleged that the 1921 agreement with Western Electric not to compete in Germany had been broken, as 'a great financially powerful American undertaking' with an 'unchecked desire for expansion . . . on the road towards the desired control of world markets'. Employees of STC were encouraged to buy shares in the corporation through salary deductions but the outlook in their part of it was very different just as their leader was not in the mould of Colonel Behn.

E. S. Byng, a Plymouth Brother dubbed 'Plymouth Rock', was a mild-mannered man who looked for good in everyone. One of his great concerns was to improve human relations in industry. To further his idea he hired consultants, the result of whose efforts was to strengthen the hand of the personnel department *vis-à-vis* foremen, who no longer had absolute rights in hiring and firing. His gestures were well-meaning but he made little impact as a businessman. Unfortunately he was managing director at a time when the country and the company were in the wake of the Wall Street crash, an event made even more dramatic by the panic the telephone helped to spread. When the no par stock was down, from a high of $149 in 1929 to a low of just over $7 in 1931 and under $3 in 1932, there was no doubt whose company it was. With his ambitions baulked for the time being, and the prisoner of economic circumstances, Behn at this juncture may have been less concerned with humanity in industry than was usual for him.

8

Slump, Invention and Recovery: 1930–1939

The effects of the 1929 crash were not immediate. STC, with sales and number of employees at a postwar peak in 1930, was still a company of opportunities. Edwin Plowden, a future chairman of Tube Investments, having come down from Cambridge in 1929 with an economics degree, was unemployed for about a year and had to depend upon his widowed mother:

> I finally got a job with International Standard Electric because my father, who died when I was a child, had been a friend of Esmé Howard, then our ambassador in Washington. I wrote to him and said I was desperately looking for a job and did he know of any American firm that had subsidiaries in England. He was a friend of Clarence Mackay, the founder of Mackay Radio, an ISEC company. As a result I got an interview at Connaught House and a job at £3 a week. They were still taking on people but they didn't know what to do with me. Masses of stuff appeared from the manufacturing companies in Europe and we were supposed to co-ordinate it.

There were, however, warning signs. In September 1930 the editorial board of *Standard News* was informed that no further issue of the monthly journal could be made that year and that in all probability the full budget of £1,350 would not be available next year. Management wanted quarterly publication. Because this would have meant using information up to four months old the editorial board recommended the publication of a monthly news sheet as well. Neither appeared. Communication of the bad news to employees was face to face. Eric Bivand was a 20-year-old at the labs:

> My boss Freddie Knight had a desk right up in one corner. He just called us up and sat us down at the side of his desk one by

> one and told us that our services were terminated. Not, he hastened to add, for any fault on our part but because the Americans had decided to cut down on the research and development effort. He was terribly upset at having to do this because he was a sort of father figure to us young chaps. He was getting near to tears, wished us all the best.

On some days there was a steady procession of people going along to corner offices and coming out a few minutes later, stunned. An ex-serviceman fainted and, actor or not, kept his job. The rest collected their personal effects and said goodbye. Morale was low. Survivors did not know how long they would last. They believed that New York cables came into headquarters with orders to book rooms at Claridges and sack so many staff. Americans were in evidence economising and reorganising, and when they had sailed away their decisions, which seemed ruthless to those affected, were implemented. Only a few hours' notice was given, generally at the end of the week, and there was money only for the next period one would have worked. Later a few early retirement pensions were granted. Some qualified people were fortunate in going to the BBC and the Post Office but others had to try their hands at starting totally different businesses, some being reduced to occupations like selling stockings door-to-door.

The situation was summed up in a single sentence in the 1931 annual report: 'The cost of conducting the Laboratories was reduced during the year under review through the temporary discontinuance of all work not of promptly remunerative character.' By the middle of 1931 all 420 employees had gone. The building was vacated and became the Police College. For fear of losing ideas to the competition, short-range developments were moved with their personnel into STC, still apparently flourishing. In July 1931 the directors were proposing a 25 per cent bonus dividend following 1930 profits of £576,224 against £422,376 the previous year. Other people were able to transfer to the security of a productive company. Plowden was one of them:

> I am quite sure the only reason they sent me to STC was because I was very cheap: £3 a week. After very inadequate training they sent me on the road to call on factories to try and sell internal telephone systems. I used to go round the trading estates knocking on windows saying 'Could I see the buyer?' More often than not there was a commissionaire who banged the window shut and said 'No'. It was frightfully difficult to

> get in to see someone. People didn't want to see you. It made me much tougher than I might otherwise have been. There was no such thing as formal sales training. You were just turned out and told to sell things. We made some sales but not many. When we did it was a triumph. I was not on commission. The incentive was to keep one's job.

In the international economic crisis of September 1931, when Ramsay MacDonald's National Government went off the gold standard, the pound was devalued by 25 per cent within a few days. By mid-December 'owing to the serious effect of the present industrial depression', the directors decided not to declare or pay a bonus dividend. People were downgraded, finding themselves working alongside those they had trained. Even if they kept their jobs, the rewards were smaller. There were three or four successive salary cuts, accepted as preferable to dismissal. In the office and on the shop floor there was no union representation to alleviate the effects of the slump. Reg Birch, a 16-year-old and the only breadwinner in his family, transferred from Hendon to Southgate:

> The bulk of the tradesmen were members of a union though they never vouchsafed it. One or two would wear a badge behind their lapel. To join I had to seek out somebody who would confess. I went to a chap Tommy Doran, who I knew for a fact was a branch secretary of the AEU, and he wouldn't admit. He said 'You'd better go and see Trotsky', a fellow who was allegedly a militant because he had a beard.

T. G. Spencer, who became managing director in 1933, later claimed to have hidden loyal servants of the company from New York. What New York did not see from afar it need not grieve about. One way to stay in work was to get sent abroad, a technique successfully adopted by Frank Wright, another future managing director, who spent most of the 1930s away from home. In the UK the policy was retrenchment. Empty offices, scattered around Aldwych, were disposed of and letting them off became a full-time job. In 1933 the idea of the grand plant at Wembley was abandoned, the Hendon plant vacated, radio activities moved to Southgate and transmission work to Woolwich. The split was dictated by economics yet was technically unfortunate. Sending signals on land, under the water or through the air involved similar techniques but their organisational separation across London from N11 to E16 accentuated the differences.

In the last days at Hendon, as space took the place of work, discipline became lax. The high old times of the 1920s were almost over. After the free and easy days of pioneering, life was going to be a constrained routine. Young engineers in particular resented moving into Southgate, the Parish of High Kipping, where they would have to endure a much stricter regime. Handwriting was now prescribed as an economic method of communication. Internal memos were not to be typed and replies were to be written on the reverse. Much draughting was eliminated by relying on rough sketches. General distribution of drawings to customers was temporarily discontinued, only essential installation drawings being issued. In some sections, particularly the export of repeater and radio equipment, the too few people were kept very busy. In addition, a top technical man could find himself doubling as an assistant to the European technical director, from 1933 Maurice Deloraine based in Paris.

In general though people were looking for work. At a time when it was postponing orders the Post Office took more unto itself, handling more of its own installation and doing more research and development. Manufacturers had to find alternative work, anything to keep going. In what some called its 'dolls' eye and hairpin' era STC tried unsuccessfully to sell German-made conveyors to Boots, made umbrella stays at Southgate, and nearly took a contract with Pathé for 1,000 cinema talkie equipments. This, however, came too near to contravening the Western Electric agreement and was passed on to Plessey. From Plessey the company bought components for the Philco wireless line it laid down at Southgate.

Its own components business also developed. The valve department was able to make small X-ray tubes, used in dental equipment. It was the most successful part of the venture into medical electronics. Some of the developments were quackery: boxes impressively fronted with knobs and dials but containing a dry battery that would do patients good by auto-suggestion; a surgical knife that cut steaks by a radio frequency burn but was never used on a human being; and somewhat better radio diathermy equipment that gave staff free treatment for rheumatics as an experiment. The beauticle was a manicuring device. In the Foots Cray, Kent, plant of Kolster-Brandes the manufacture of domestic radio valves under the brand name Brimar was established. As electric companies changed from supplying d.c. to a.c., there was a demand for power-conversion units that could be used with existing d.c. wireless sets. This led to the importation from an

ITT company in Germany of selenium rectifiers, used mainly by the makers of battery chargers.

A whole new business developed from the supply of a few relays and cables to the Bell Punch Company, which among other things made totalisators, then being installed at greyhound stadiums. Soon the company was working with Union Totalisator installing totes at home in places like Brighton and Cardiff and abroad at race-courses including Accra, Baghdad and Hong Kong. A mobile unit was made for New Zealand. By the time war broke out in 1939 the entrepreneurial Reg Ballard, who had made this a very profitable business and was planning an attack on the US market, was also going into photofinish, using American technology. The Jockey Club did not want to know because it might suggest that their judges could make a mistake. Dog tracks like Wembley were quicker to take it up. The tough years were a spur to enterprise but the results were not immediate. In 1931 net profits plummeted to £36,163 against a 1930 high of £576,224. A loss of £50,277 was recorded for 1932 and in 1933 a profit of £78,945 was struck only after adjustments. Without these there would have been a loss of £44,313.

In the circumstances the well-meaning Byng did not command the respect of the employees. They had much more sympathy for his younger brother Freddie, who liked to get abroad and kick over the Plymouth Brethren traces. As a face-saver E. S. Byng was made joint managing director in 1933, an arrangement that lasted for two years, when he was elevated to vice chairman. Frank Gill, widely regarded as one of God's own gentlemen, was made chairman.

The man who was to give the company its character for a quarter of a century was the managing director Tommy Spencer, a Woolwich cable man who had a good memory for names and a bluff ability to get on with people at all levels. His employees believed that, because he was so popular with their major customer, he had his own in/out board at Post Office headquarters and could get payment in advance to meet the payroll. Privately, he could be mean and mean-spirited but he could project the degree of East End mateyness right for an occasion. It made up for any limitations he had as an engineer or businessman. He was able to surround himself with a loyal workforce yet keep his degrees of distance from them. A former half back for Woolwich Arsenal, he encouraged sport within the company and saw inter-location tournaments as a means of raising the depressed morale and

1 Western Electric upright manual switchboards in use at the Holborn telephone exchange, c. 1900.

2 Western Electric employees in the metal plating shop, North Woolwich, before the 1914–18 War.

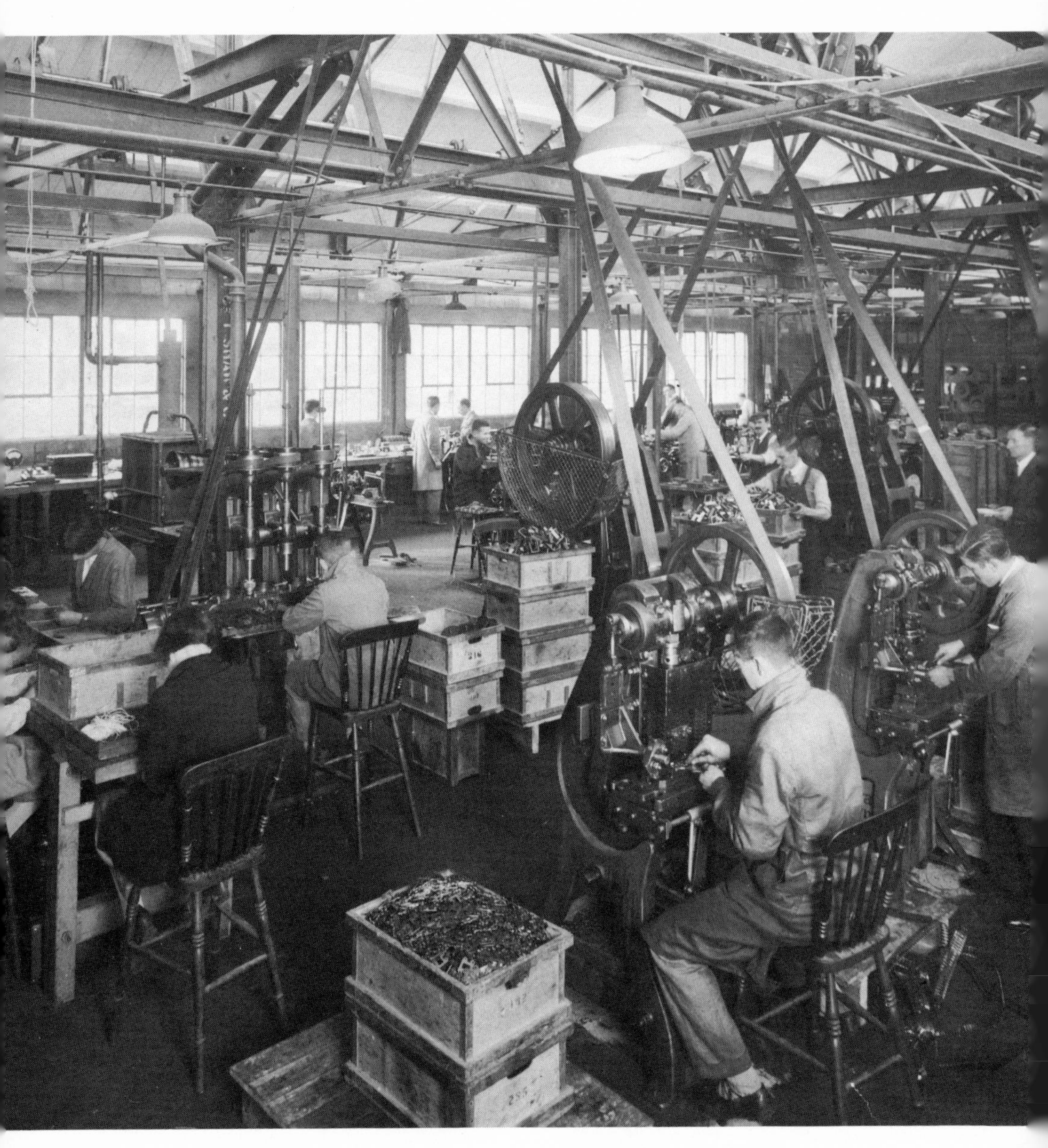

3 Belt-driven machinery and makeshift conditions were common in many British factories right up to the 1950s. This picture shows the piece-part shop at North Woolwich in the late 1920s.

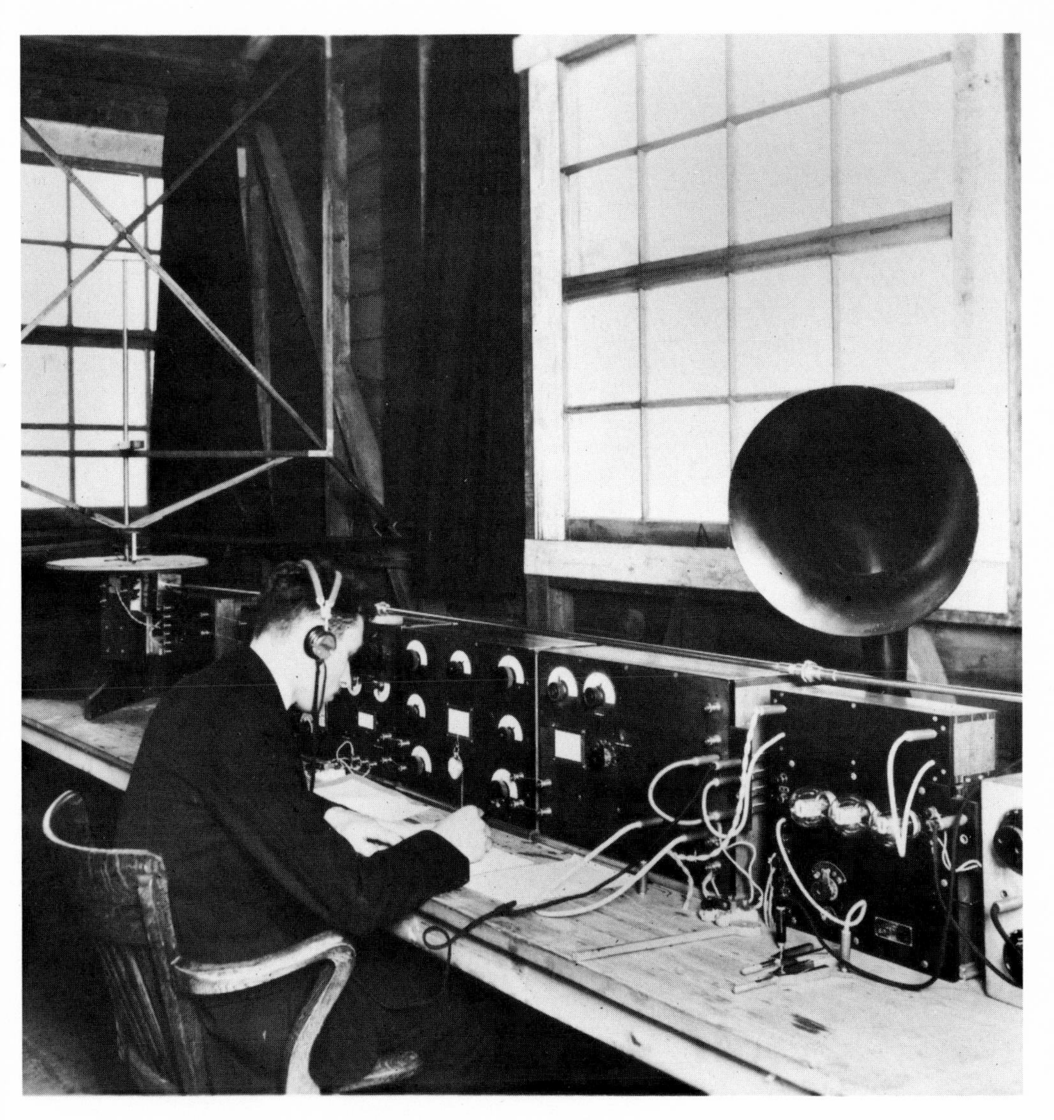

4 Dr Maurice Deloraine, later technical director of ITT, tests receiving equipment in preparation for the first one-way transatlantic telephone call, made between New York and the New Southgate factory on 15 January 1923.

5 Wireless sets are popular with Bright Young Things in this Western Electric advertisement of 1924.

6 One of the souvenirs produced for the British Empire Exhibition of 1924 by Western Electric was a set of Mah Jong cards showing a selection of company products.

7 Postmaster-General Sir Kingsley Wood reviews a Post Office promotion at GPO Headquarters, London, in 1932.

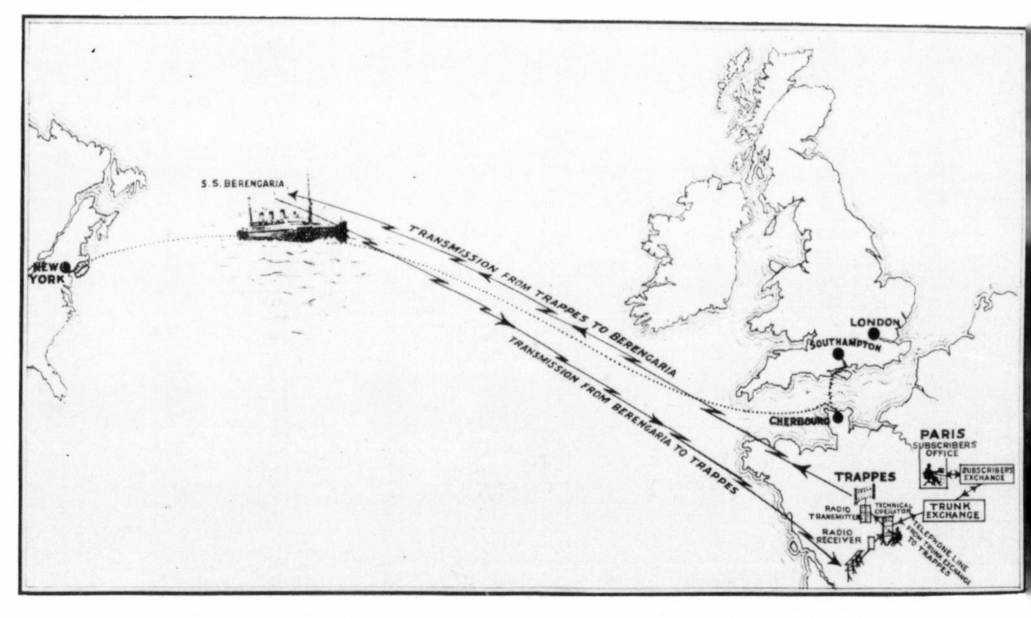

8 The first ship-to-shore telephone call over the public telephone network was made from the *Berengaria* to Trappes, France, using STC equipment, in 1929.

9 An STC public address system in use at the London, Midland and Scottish Railway shareholders' meeting in the Great Hall, Euston station, 1928.

10 The light and space of STC's New Southgate factory contrast with conditions at North Woolwich (see Plate 3). Here, PBX switchboards are made in surroundings that, in 1930, were far in advance of those prevailing in much of the industry.

11 The world's first commercial microwave link, using the 'micro-ray system' demonstrated by STC and its sister company LCT in 1931, was put into service on 26 January 1934 between aerodromes at Lympne, Kent, and St Inglevert, Pas de Calais, France. This photograph shows the microwave dish at St Inglevert.

12 International Marine Radio Company operators at work in the control room aboard the *Queen Mary* on her maiden voyage in 1936.

bringing together the organisation he had inherited. It was fragmented both by its geographical dispersal and by the broadening range of its technology. This was recognised in a reorganisation based on product lines. People in different product lines on the same site had little in common. They were more likely to be looking over their shoulders at their competitors than towards their colleagues next door.

Changes had been taking place in the Post Office. In a 1929 pamphlet *The Stranglehold on our Telephones: a Practicable Remedy*, the Telephone Development Association argued that among other things:

> Our existing system of telephone administration is fundamentally wrong . . . What is wrong is the substitution of political control for business control . . . The Post Office is a department of the Civil Service, and as such, is wholly staffed by Civil Servants . . . But the technique of the Civil Service is not necessarily well-suited to the conduct of what is essentially a business undertaking . . . It is quite wrong that the control of policy in a vast industrial undertaking (which is what the telephones have become) should be vested in a Minister . . .

What the Telephone Development Association wanted was a public corporation, like the BBC, responsible for telephones and not dependent on the Treasury for finance. It pressed its case in a relentless campaign, including letters to the prime minister and articles in *The Times*.

As a result, in 1932 the government appointed a committee under Lord Bridgeman to look into the status and organisation of the Post Office. Its recommendations were more concerned with internal reorganisation than the importance of telecommunications in the life of the nation. Treasury control of finance remained, telephones and telegraphs were not to be separated from other Post Office services, and no new public corporation was to be established. Such administrative changes as were made did improve day-to-day working relationships with the industry but much more depended upon the energetic personality of the new Postmaster-General, Sir Kingsley Wood, though he was unable to do anything immediate to help the industry over the slump. That depended upon general economic recovery and specific technical progress. An outcome of the new relationship between the Post Office and the industry was the formation in 1933 of a joint committee, British Telephone Technical Development Committee, to promote detailed standardisation of equipment. A complementary committee was set up by the manufacturers to implement standards and retrospective changes.

In 1930 there were almost 2 million telephone subscribers, by 1939 3·3 million. Growth accelerated throughout the decade. Industry and commerce were concentrating in larger units and the telephone was an essential instrument in running an enterprise like a chain of stores. With a few exceptions like Evelyn Waugh, professional people found it useful in carrying on a practice. Individual subscribers were attracted by continually falling charges. From 1936, the year that the speaking clock TIM was introduced in London, there was no timing on calls up to fifteen miles and people could talk locally forever for a penny or twopence (less than 1p) in a call box. Moreover, residential subscribers got fifty penny units free every quarter. Helping to reduce costs were technical advances making it possible to transmit a greater concentration of signals over a given route.

The most spectacular, spectacular enough to appeal to Behn, was a new technique of communication. Radio engineers working under André Clavier in Paris had been experimenting in the production and reception of extremely short waves, about 17 centimetres against the previous record low of a metre. These wavelengths could be concentrated in a sharp beam by a 'dish', a parabolic reflector about 3 metres in diameter, and the highly directional beam projected on a line of sight to an identical 'dish'. Preliminary tests suggested that the Channel could be crossed at its narrowest point, using almost the same sites as Blériot on his 1909 first crossing by air.

On a cold 31 March 1931, 'dishes' on the cliffs above St Margaret's Bay near Dover, operating on a wavelength of 17·6 centimetres, sent and received telephone messages from Calais. Reception was up to the standard of a high-quality cable circuit and there was little evidence of fading. Further possibilities of the new system were demonstrated with the sending of printed text by facsimile, another novel equipment. Its speed of about one page per minute was faster than existing telegraph systems. Guests on both sides of the Channel were impressed and STC and its associate company in Paris, Le Materiel Téléphonique, had every reason to be pleased with the demonstration. Responsible for the press arrangements was Bernard Holding, an ex-editor of *The Electrician* and now publications manager for ISEC. To describe the new technique he coined the words 'micro-ray' and 'microwave'.

Hitherto, the accepted view had been that the spectrum of frequencies available for transmission was strictly limited and hence that bandwidths had to be conserved. Now it was possible to develop economically virgin territory, bandwidths for thousands and thousands of channels. In a note explaining the significance of the development STC pointed out:

> Further very important use will be for television, the development of which is hampered at the present time by the very large frequency range required for satisfactory definition of the object transmitted. It should now be possible to allocate as wide a band as is necessary for television without causing any ether congestion. It is easy to imagine the establishment of national 'micro-ray' networks for use in conjunction with television apparatus.

This was in the year following Baird's experimental public service and five years before the opening of the regular BBC service from Alexandra Palace. The most extensive use was to be postwar.

The immediate practical application of the new technique was for the British and French air ministries as a link in their direct air traffic control circuit between London and Paris. In 1934 the world's first microwave radio service was opened between the civil airports at Lympne in Kent and St Inglevert in France, thirty-five miles away. Teleprinter messages, at a speed of sixty to seventy words a minute, and telephone conversations could be transmitted simultaneously. During installation, to get the siting right, the station at Lympne was fixed and a lorry carrying the other 'dish' was driven up and down at St Inglevert to arrive at the appropriate direction and then height.

At this period everybody was so absorbed in the exciting advantages of microwaves as a means of communication that nobody realised the significance of another phenomenon. Clavier had observed that when a ship in mid-Channel interfered with the line-of-sight transmission the position and size of the vessel could be determined. An English radio engineer, Alec Reeves, out in a rowing boat, was aware of signals being reflected from the cliff face. Had they pursued these observations, they could have developed a radio detection and ranging system, radar.

A few developments of the Hendon labs became commercial. One was the eighteen-channel voice frequency telegraph system, on which Leslie Long did some of the early work:

> We lost out in the arguments in the CCITT to Siemens & Halske and we hurriedly had to change our frequency spacing from 170 cycles, which was copied from the Americans, to 120. We did it and got the whole of the Post Office business. Nobody else had a look-in.

It was used by the Post Office for replacing practically the whole of its long-distance telegraphs, the first being between London and Dundee in 1932, followed by London–Glasgow–Belfast. Another development in transmission taken up by the Post Office was the application of carrier techniques to telephone cables instead of overhead lines. The first system, with twelve channels, was installed in 1936 to increase the number of circuits between Bristol and Plymouth. R. A. Mack, the engineer responsible, had visited Bell Labs in the USA, proceeded with his design and beaten them on the first installation. It became the premier means of providing more circuits to cope with the growing long-distance traffic in the UK.

Increasing the number of channels on existing routes, as the Post Office wanted to do for economy reasons, caused problems like cross-talk, especially on older cables. A new design of higher-capacity, smaller, lighter cable was needed to match the advances in transmission techniques. The answer was a coaxial cable, so called because it consisted of a centre conductor separated by an insulator from an outer conductor, the two conductors being concentric. There was nothing new about the basic design, which the physicist Lord Rayleigh had sketched out in the late ninteenth century. It remained an academic theory until there were advances in materials. To make it a reality all that was needed was a near-perfect conductor and, more important, a near-perfect insulator. For this, STC experimented with polystyrene, ebonite and an acetylated cotton yarn.

The textile worked well enough for it to be used on the first coaxial cable made and installed in Britain, providing in 1936 forty channels between London and Birmingham, the route on which the company had introduced loading coils nearly thirty years before. Already it was being referred to as 'television cable'. By 1939 the route was extended to Manchester and then via Leeds to Newcastle, the final link being an experiment in providing no fewer than 600 channels. As coaxial technology developed, the insulator consisted mainly of air with polythene discs at 1·3 inch intervals separating the conductors, and copper tapes of a high degree of purity and accuracy also had to be produced. The construction was apparently simple but as with other cables the secret of success was in the production processes.

One concept was destined to remain impractical until an advance was made on the valve. Its inspiration was pre-Bell. After Morse invented his telegraph in 1844 others tried to translate speech into a code suitable for transmission but it could not then be done. Bell worked on a different concept, the analogue principle that the variations in the human voice produced a proportionally varying electric current. Trying once again to digitise speech was the idea of Alec Reeves, always a less than conventional radio engineer. He had worked at Southgate on the transatlantic propagation tests leading up to the Rugby transmitter, invented the condenser microphone instead of the carbon type for broadcasting, and missed the invention of the superheterodyne receiver by a fortnight. In the Paris labs he worked on the application to short wave of single-sideband transmission, a technique that had the advantages of reducing the necessary transmitter power and bandwidth. It was successfully demonstrated in 1931 on high-frequency radio circuits Paris–Madrid–Buenos Aires but the slump was not the ideal time to apply it. Like most of Reeves's work it had to wait for the world to catch up. During and after the Second World War, when it was adopted by the Post Office and AT&T, it became the universal system for high frequency point-to-point telephony.

After observing microwave signals being reflected from the white cliffs of Dover, Reeves had more and more become convinced that people in other universes were trying to communicate with us and that their signals could be translated into morse code. He began to log his interpretations in a red exercise book, his strongest contact being with the great electrical pioneer Michael Faraday, whom he looked upon as a friend and adviser. Sometimes he sought information from Red Cloud. A brilliant individualist, Reeves lived in two worlds: dabbling in spiritualism, trying to comprehend our sixth sense; and the practical world of the company radio lab. Having been told that dealing with the paranormal entailed risks and that some protection could be obtained from gold and steel in water, he later kept at the back of his recorder a glass of water containing a half-inch drill and the OBE awarded for his war work.

In 1935 a nine-channel radio link was demonstrated between Cap Gris Nez and St Margaret's Bay. As a result, the Post Office ordered a similar system that went into service between Stranraer in Scotland and Belfast in 1937, on which, with the problems in preventing intermodulation between channels, speech could suddenly sound like a mixture of Scottish and Irish bagpipes. Reeves looked at various methods of transmission that could overcome the problems of noise, distortion and cross-talk. In 1938 he filed a patent in France that changed the existing basic concept. His idea, pulse code modulation, was to substitute digital coded signals for the analogue signals used for more than sixty years since Bell's invention of the telephone.

The British patent was filed in STC's name in 1939, the month after war broke out. Reeves returned to England in June 1940 but there was more urgent work for him to do. In December 1940 Deloraine, who had worked with him, went to the USA, where the military showed interest. An experimental order was placed with Bell Labs, a full-scale, twenty-four channel system was used in the field by the US military at the end of the Second World War, and postwar the Bell system was to benefit first from the new technique. Because cabinets upon cabinets of valves were required, usage did not become general until the advent of reliable transistors and integrated circuits, by which time the basic patents had expired. STC made no money from them. Fittingly, the idea that came from outer space had an application back there. It was to make possible the transmission of pictures from space satellites and the moon to earth.

Domestic radio had also been advancing. In 1930 there were just over 3 million wireless licences in the UK; by the end of the decade nearly 9 million. Expansion on this scale meant good business in microphones, valves – both receiving and broadcast – and transmitters. Indeed, those working on transmitters were almost too busy to notice the lack of work in the world outside. Relations with the BBC were in many instances personal, with former colleagues who had left or had to leave in the slump and had joined the Establishment. On both sides, as they worked on higher-power transmitters to reach farther distances, there was a lively feeling of being pioneers. In 1932 the first short-wave Empire broadcasters were cut into service at Daventry, just in time to carry the first ever King's Christmas message to his scattered subjects. Within the company there was elation that they had beaten the Marconi company to the order.

At Daventry two 80-kilowatt transmitters were commissioned in time for the coronation of King George VI on 12

May 1937 in Westminster Abbey, for which the company also supplied the public address equipment. Again the STC people patted themselves on the back – the Marconi transmitter was not ready. Two 100-kilowatt equipments were supplied for the regional service and a fourth for Daventry. Tests in various parts of the world showed that 'it is now only on rare occasions of exceptionally disturbed ionospheric conditions that it is impossible to receive news bulletins from Daventry, on a reasonably good commercial receiver, in any part of the world at least once in every 24 hours'. Of more UK interest was the transmitter – Marconi also supplied one – for the experimental TV broadcasts of the Baird and EMI systems from Alexandra Palace from 1936.

At sea, radio was coming into its own. In 1925 the company equipped an Antarctic whaling fleet and its base station in South Georgia. This was believed to be the first time that radio-telephony was used commercially for communication between ship and ship and ship-to-shore. It was so successful that whale catches had to be restricted. To compete with Marconi International Marine, Behn set up the International Marine Radio Company in Connaught House in 1930, following the first telephone conversation ever held between the normal public telephone service on land and a ship at sea, the *Berengaria*. The company's first notable contract, including the supply of radio officers, was for the same Cunard passenger liner, yet its formation was more competition for competition's sake than an appraisal of the market at a difficult time. Some small installations were made but the first year's trading was disappointing and the company became dormant until new management started to revive it in 1932. Later liners equipped included the *Aquitania*, *Majestic*, *Mauretania*, *Olympic*, *Queen Mary* and *Queen Elizabeth*. For the two *Queens* the whole of the radio equipment was supplied.

Whereas the League of Nations was revealed more and more as a body without power, its humbler counterpart in telecommunications had a growing strength. Not a supranational body, its thoroughly discussed recommendations in telegraphy, telephony and radio were acquiring the force of standards, standards that were largely implemented by European manufacturers. They were able to sell products and know-how to an undivided world. At the beginning of 1927 no telephone subscriber could speak over more than 100 miles of water. With the advances in radio, by the end of 1931 subscribers in the two Americas, Europe and South Africa were in actual or potential contact with one another and people at sea were not isolated. For STC, with a broadening range of relevant products, the decade was the peak of exports in terms of geographical distribution.

In 1932 a high power – 120-kilowatt – broadcaster was installed in Prague, just a few months before a similar Marconi installation in Warsaw and the forerunner of a number of prewar stations with powers up to 140 kilowatts. Installations with these outputs stimulated developments in valves and associated equipments, where mercury arc rectifiers helped solve the problem of instantaneous switching off and on, which could blow the expensive valves. Budapest boasted a 1,005-foot mast, the highest structure in Europe until it was toppled in the war. Business in Eastern Europe for cables and transmission systems was helped by the existence of local ITT companies and by ITT assistance to telephone administrations shrewdly negotiated by Behn at diplomatic dinners. His employees followed in his wake on the *trains deluxe* or frail low-flying aircraft, bringing with them repeater and carrier technology. There was also work in the Empire, reorganising and extending systems. Having successfully adapted Western Electric mass production techniques for a smaller cable factory, STC was able to start securing a steady royalty income through sub-licensing overseas.

Radio was a relatively small but growing business in aviation. In 1931 the company produced its first airborne communication sets. These were in fabric-coated plywood boxes suspended in aircraft by loops of catapult elastic to stop the valves shaking out of their sockets. Remote control was by levers and wires. In 1936 this form of construction gave way to a metal chassis with an anti-vibration mounting. It was a private venture designed originally for fighter aircraft. Radio was also a means of navigation. A beam approach system developed by an associate company in Germany, Lorenz, was installed at the three main London airports: Croydon, Gatwick and Heston. Charles Strong brought a radio altimeter back from the US and development work started on that: 'We planned to use that particular technique for doing radar. That was a bit of a mistake on our part but the government did not at that time put us into the full radar picture because of our international associations.'

Internationally the company was trying to develop its aviation business with practical demonstrations. To back them up L. J. I. Nickels was writing tenders in the radio division:

> H. M. Samuelson, who was deputy head of the commercial department in Connaught House, got himself a pilot's ticket and persuaded the company to invest in a Tiger Moth. We moved on to a Puss Moth and eventually a Heston Phoenix, a wooden aeroplane, nominally a five-seater, grossly under-powered. It had to have a proper pilot not a half-timer and we got a young Dutchman, Albert Oping. He flew this thing all round Europe. In Bucharest he landed it with the wheels up and again in Riga on the Baltic States exploitation tour. That bent the propeller and it took some weeks to get a new one from London just before the war.

The most ambitious export contract taken was where the company had no political pull, in the Soviet Union. While memories of the 1933 arrest of six Metrovic engineers by the OGPU on charges of sabotage and bribery were still fresh, the company started negotiations on providing the carrier system on the longest telephone line in the world, the 6,000 miles from Moscow to Khabarovsk. Leslie Long was the engineer of the two negotiators:

> The sales people decided that we were so hard up for business we could go and look anywhere. We were talking to the Russian Railways first of all and got some orders for three-channel carrier but we'd also got involved while we were in Moscow on the Moscow–Khabarovsk scheme for the administration. We came back and started making some definite plans about it at the time when the Russians were hand in glove with the Germans, Siemens & Halske, and they were getting a bit fed-up. The political situation between Russia and Germany was getting difficult and we stepped in. The Russians were very enthusiastic, very persistent, talk and talk and talk. To get a decision you had to say you were going home and then things began to happen.

There were qualms within the company about undertaking a project of such magnitude. It was twice the length of anything covered by a similar system in the USA. Tommy Spencer called a large meeting of the involved people and put it to an open vote. They decided to go ahead and the three-channel carrier was made and delivered just before war broke out. Soviet engineers installed it. Forty years later a delegation visiting the UK said that it was still working but a few old timers were not allowed to retire because they were the only people who knew how to maintain it.

At home, to cope with the economic recovery from 1934

more buildings were erected on existing sites. At Woolwich, beginning with a modest store in 1934, over 200,000 square feet were added by 1940. New products accommodated were quartz crystals and rubber-covered cable. Woolwich was already the crowded home of cable, transmission equipments, microwave and components like condensers and magnetic materials.

At Southgate, which had exchanges and radio as its two basic products, space was more ample. Further land was acquired in 1933, 1934 and 1939. In 1933–4 a third storey was added to the original mill-type structure and further building on the site more than doubled its manufacturing space to over 250,000 square feet during the decade. Towards the end of the period, with Air Ministry support, work started on a new building for the envisaged wartime expansion of radio communications. In exchanges the emphasis was on flowline production, with parts moving on conveyors to assembly benches and on for rack mounting. Norman Kipping set out to make it an efficient showpiece plant. On 8 February 1935 it merited a visit from the president of the Industrial Welfare Society, the Duke of York, soon to be King George VI. He planted a flowering cherry and received two ivory and gold-plated telephones for his daughters. Not to be outdone, Colonel Behn planted another flowering cherry on 13 August. His lived.

In 1938 the company got back into the business of wireless sets with the acquisition of Kolster-Brandes and with it a plant at Foots Cray in Kent. There were mixed feelings in the company whether it was the sort of business it should be in. It all added to growth. Between 1928 and the outbreak of war the number of employees increased by 78 per cent and business by 112 per cent. From ambitious schemes, through depression and uncertainty, it emerged as a period of solid achievement. People had been chastened by the Depression. Bright young things of the 1920s had grown up into a more serious age in which it was important to hold down a job and do something. In the 1930s life was more purposeful, perhaps duller, but there was a bit more to show for one's efforts. Although the labs had been closed, development had not ceased and was providing present business, while telecommunications techniques were fanning out into new areas that would provide future work. In the meantime an increasing portion of work in hand was coming from the rearmament programme.

9

A Company of Gentlemen: The 1930s

STC was mainly a light industry in new technologies with growth prospects. With all its three plants in the London telephone area, it was firmly in the most prosperous part of the country, south-east England. As the 1930s progressed the company took back some of the employees it had dispensed with. One of them was Eric Bivand, who after being dismissed from the Hendon labs had joined Ferranti and been exiled to Lancashire:

> I had a girlfriend in London and I used to come down as often as I could, which was not very often. She took a poor view of this and so did I so I decided that I would try to get a job in London back with STC. I wrote and eventually got a letter asking me if I could attend an interview with E. A. Richards at Southgate. This posed a problem because the interview was on a Wednesday and I daren't let Ferranti know that I was applying for another job. I would have got the sack there and then. I phoned up Richards, who very kindly offered to see me on a Sunday morning at his home in Stanmore. He was a charming typical gentleman. When I got to his house there were two milk bottles outside and no sign of life anywhere. I knocked and eventually he appeared in a dressing-gown, terribly sorry but he had forgotten I was coming. Would I go into the sitting-room and read some books while he dressed. On the table was a grey book about selenium rectifiers. I had never heard of such a thing. While he was dressing I read the book from cover to cover. He was the author and it was the first publication on the subject. During the interview he was terribly impressed with my answers. I got the job. My letter of employment dated 28 January 1938 began 'We have now heard from your present employers that they are prepared to release your services and that you will be able to commence duties with us . . .'

Ken Frost, who was to rise to hold several senior appointments in the company's telephone and exchange business, joined for two reasons:

> It was primarily an American company. I thought it was going to be more progressive and more active. Secondly, I was finding the going very slow at Faraday House, where I was taking my degree, so I joined STC as a sandwich student, taking a degree course at Woolwich Polytechnic.

There were also able engineers to look up to, men recruited by the technical director, A. W. Montgomery, whose work in their various fields had led to new techniques and products that helped to lift the company out of the Depression and promised well for the future. There was much to learn but, although it was a good training ground, Frost did not find it particularly 'American and efficient . . . It was very much a group of divisions and a group of people working together in a division. There wasn't really a company. Connaught House was a remote place.'

F. C. Wright was head of the carrier planning group at Woolwich:

> One wasn't even allowed to know what proportion of the company's capital was being allocated to one's particular application . . . I am quite sure the switching people thought that transmission people were absolutely the end and no cable engineer ever thought of speaking to me. It would have been social death. It really was a conglomeration of tribes living in a state of armed neutrality. Loyalty to the company was never undermined by this but it was just unfortunate that we had such unpleasant neighbours who also had the same sort of loyalty. You did your best in the circumstances not to be too unpleasant to them because one was to a degree interdependent.

As if the divisions of geography and technology were not enough, the company added an organisational split. On one side was manufacturing, on the other sales and engineering. Inevitably there was conflict. It was not always possible to make what had been designed or sell what could be made. Equipment engineers were not allowed to see sales quotations or encouraged to produce cheaper designs for manufacture. Differences were compounded where responsibilities were further divided. Cable department at Woolwich was responsible for the manufacture of transmission equipment, for which some of the parts were made at Southgate at high cost. Transmission felt like the poor relations. The whole cumbersome process slowed down their pace of development.

When it came to the resolution of conflict the arguments were stacked in favour of manufacturing. It had a plant to keep going and what it was tooled up for was a steady run of work with the minimum of change. The bulk of the effort was in making standard items for the Post Office. What the major customer wanted was uniformity and bureaucratic procedures were developed within manufacturing to ensure it. If prototypes, one-offs and specials had to be produced there were model shops within the company. Moreover, the person responsible for a site, the works manager, was a production man. To make matters worse there was enmity between Kipping at Southgate and Pheazey at Woolwich, the difference later being emphasised by Kipping's knighthood against Pheazey's CBE.

In Southgate, Kipping had something to be proud of. Reg Birch, a young Marxist toolmaker, could also take pride in part of the capitalist system:

> New Southgate was a dream for someone trying to learn his trade. At the end of a shift when you walked through from the toolroom the press shop was by industry standards enormous. The favourite thing was to go and see if your tool was up and punching all right. The strip was on a roll going through automatically for stamping – permalloy, laminated natural phenol fibres, ebonite, nickel silver, brass, copper, phosphor bronze, bright steels, cold rolled, close annealed – all being punched up and there would be a light on the press. It was like Ali Baba's cave. The pieces would be falling down the hoppers. It was a myriad of colours, almost like base metal, gold, silver, platinum. It was romantic. It was a pride and joy for me to see. I made some of those lamination tools. They made thousands of transformers so they made millions of laminations.

People prized their jobs, kept their heads down and got on with the work. There was a strong sense of duty, partly engendered by the knowledge that there was somebody at the gate ready to step into your place. People were careful about time-keeping. Too many warnings for lateness and you could be out. Not that the company was a great payer. When recovery from the Depression was starting, a 1½d (just over ½p) an hour shop floor increase was spread over six months:

¾d in the first three months and then ¼d a month for the next three months. A sandwich student got £1 5s (£1. 25) a week, a chap in his 20s with experience £4–£4 10s (£4. 50), and the £5 a week man was somebody. In theory nobody knew what anybody else was paid and if you got a rise you were told not to tell anybody. A half a crown (12½p) rise had to be approved by New York. Only for the select few were there substantial rises and they were in special circumstances. When at the end of 1936 Francis McLean decided to go to the BBC, Spencer offered him the massive rise of 50 per cent to stay but McLean's mind was made up. Had 25 per cent been given earlier, he would not have thought of moving.

When weighing up a move employees had to take into account their pension benefits. With a non-contributory scheme there was no refund of contributions on leaving and all pension rights were lost. It helped to bind employees to the company but some felt that in the long run it did a disservice, acting like a filter that let the brightest people out and retained the less able and ambitious. Having such a scheme was unusual but it sometimes seemed more to the advantage of the employer than the employee. The pension scheme was an argument for paying less, an argument that personnel officers traded on. Recruits were offered a security. To the company too the scheme was a regular source of funds. Contributions were paid over to Midland Bank Trustees, which as part of its investment lent back some of the money. Indeed, after the Depression it helped the company stay afloat. It was not necessary either to reduce capital or raise more by going public. Both ideas were examined and discarded.

In these hard times the company had one car, a Daimler with straps for visitors' luggage. It came under the secretary's department and there was a strict hierarchy as to who could use it. With resources scarce, people watched pennies, even to the point of inefficiency. Joan Smith joined the patent department as a £1 a week junior in 1936:

> We were not supposed to have private telephone calls, except under special circumstances. If we needed to get an urgent message to our home we had to ask. You'd probably get permission. There was only one telephone between about 12 people. That was on the supervisor's desk. None of the girls used the phone for normal business. A distinction was made between a telephone call and a Post Office call i.e. one on outside lines. The company itself was not a great user of the telephone to speed its own business.

Dictation was on to wax cylinders, which when transcribed were shaved so that they could be re-used. Some skills, like those of young Joan Smith, were not used:

> If you joined at 14 you could not do shorthand typing. Because there were so many older people waiting to do it you had to wait your turn. It was forbidden before you were 16.

There was a very definite hierarchy:

> The few at the top had personal secretaries and even took them abroad with them. On the Continent not so much English was spoken so a man might feel it necessary to have his own secretary with him. If you were a senior secretary you were quite often asked to go abroad.
>
> Then there was quite a big gap. Those in between had to work for several people. For six to eight weeks I worked for about eight people in the patent department when somebody was off sick. You were not called a secretary but an assigned stenographer. So if a man wanted some work done there was a particular person to do it for him.
>
> There was quite a lot of snobbery among the secretaries about which of the other female staff they became friendly with and went to lunch with.

A girl who struck lucky was Ena Henderson. One day when they were short of secretaries she was sent up from the typing pool to take dictation from Colonel Behn. He immediately took a liking to her and she became his personal secretary, travelled with him, married the head of the Cuban Telephone Company and eventually retired to Florida.

Office discipline for Miss Smith was strict:

> You had to report on time. It was soon noticed if you weren't in on time. Limited lunchtime. I was even given an appointed time to go to the cloakroom morning and afternoon. In the patent department, which was a large department, they didn't want everybody to go missing to the cloakroom at the same time. People accepted it . . . Until the war no married women were allowed to work in the office . . . Nobody was allowed to wear very short skirts showing the knees. It was forbidden to go without stockings until the war came, when the firm couldn't very well insist . . . Sometimes the men wore sports jackets and flannels on a Saturday morning.

But golf or tennis gear could not be taken into the office. It was discreetly deposited at a station left luggage office and a sporting cloth cap concealed in a briefcase.

At Woolwich, where anybody who was anybody wore pin-striped trousers and a black jacket, you stood to attention when you went up to see the boss. At Southgate one Saturday morning in 1938 a girl who arrived in slacks was sent home for being improperly dressed because, although 40 per cent of the work force there was female, the relaxation in dress rules applied only to men. Married women were allowed to work in factories but not in the same department as spouses. A particularly irritating regulation at Southgate was that personnel in offices were not to smoke between 9.30 a.m. and 4.30 p.m. With the appointment of R. A. Miles as personnel manager for the company and as money became available, however, physical conditions were improving.

At Woolwich, Jack Davis was pleased with the new canteen and much more besides:

> Compared with the last two canteens it was an absolute palace. Things were improving all round. The dirty iron washing troughs were replaced by white china ones. There were also paper towels and liquid soap. The toilet cubicles were now properly partitioned with doors, and nice cloakrooms were being put in . . . I had discarded my big cap, white choker and red tommy [food] handkerchief and also my hobnail boots. It was now no cap, collar and tie, shoes and wickerwork food basket. I still used a tea can but this one was white enamel with a lid that was used as a cup.

At Southgate what had been the Alderman Café at the 1924 Wembley Exhibition was re-erected from Hendon, and was an early example of a self-service cafeteria. A restaurant, new kitchens and servery were added just before the war. Management had its own dining-room.

Dining arrangements reflected the stratified society. At the top was the managing director, Spencer, who alone communicated with Behn. He was the personification of the company to both New York and the employees. Although he joked with his top men he maintained a suitable distance, addressing them by surname. Below them were the monthly staff, the elite of employees, and the weekly staff. The hourly-paid punched the clock and were occasionally honoured by a photograph in *Standard News*. One of them likened the organisation to 'the railway in its stability, only it was a freer form of benevolence'. In times of hunger marches, the means test and the 'murder' of Jarrow, STC employees cycling to work knew that they were fortunate. They felt they owed a lot to Tommy Spencer, a shop-floor man who had made it.

For his part he played the paternal card well. Stories about him were many and grew in the telling, which he did not discourage. To employees he was both the remote leader occupied with higher policy and the enthusiastic sportsman who could cheer on his team from the touchline. No matter if he stopped on his way through the shops to tell a man to take his hands out of his pockets. The next minute he'd have a bluff remark for the tea-lady and everybody was grateful for his special relationship with the Post Office. He had guided them out of the Depression and was expanding the company, creating work and better working conditions.

Occasionally people had qualms about automatic devices being put on some machines and the introduction of labour-saving appliances like fork-lift trucks. They made life easier but would they also add to the problem of unemployment? The possibility was always there, though when it came down to it you were better off inside than out. You only had to look at the plight of people on the streets or at some of the rat-shops in the engineering industry to see how well off you were. Woolwich might not have been such a marvellous site physically but it was a very friendly place. Semi-rural Southgate had trees and at weekends, as at Foots Cray, employees were happy to come to the site to play cricket and other sports.

Even rebels were treated benevolently. Twenty-year-old Reg Birch, who postwar would stand out as the Maoist of the trade union movement, was angered by the Japanese who had made white men in Shanghai take their trousers off:

> STC had a visit from a Japanese trade delegation at the time the trade unions were vociferous on a Japanese boycott. I got the apprentices to stage a demonstration as the Japanese walked through. Normally we kept our heads down when there were visitors and perhaps giggled at their strange dress. The Japanese in their pin stripes came through. We all made bangs, raspberries etc. I had turned a mouthpiece on a lathe and put it in a piece of copper tubing. It made a beautiful raspberry.
>
> The supervisor came up to me with another chap shaking and said: 'We think we really ought to part company.' They were quite gentlemanly. I said: 'Why is that?' That was disaster for me, my mum, everyone. The supervisor lost his temper: 'I'm fed up with you. You're a bloody nuisance. Always

mucking about making trouble.'

I went up to personnel. The personnel young fellas were very very propitiating. They said: 'You have been a good worker here. We shall give you an excellent character. Will you come back in a year and we will give you a job?' They were as good as their word because when I stood in the Old Bailey early in the war on an industrial matter – it was not criminal but a strike against the law – my previous character was read out. This ginger copper stood up and read out one reference after the other. The STC one was 'a diligent, honest craftsman of the first order'. I would never have known but for that trial.

Reg Birch had no difficulty in getting another job, for having been a tradesman at STC was an unquestioned reference. If his sons had been so inclined he would have tried to persuade them to be apprentices there. At all levels it was a good company in which to train and you could stay and wait for dead men's shoes. Once you were in 'The Standard' you could feel you had a job for life. In the circumstances it was not surprising that there had been little enthusiasm for trade unions on either side since 1926. Wage rates in the industry were largely determined by the Engineering Employers' Federation – a paid week's holiday, granted by the company in 1930 to hourly-paid employees with ten years' service, was given to all in 1937 – and local bargaining was concerned with piece-work rates and other domestic issues. Active trade unionism was not likely to change anything fundamental.

Everybody in the company was living in a protected world. Free Trade had died with the slump but the telecommunications industry had anticipated the change of economic circumstances and philosophy. In its protected home market competition was restricted so that equipment could be standardised and the only customer that mattered was a party to the agreement. In international markets there were agreements between established companies defining territorial and technical rights. Their purpose was the sharing of available business to mutual advantage. Those at a disadvantage were the outsiders, enviously wanting to share in the assured and not unprofitable business.

In Britain the privileged suppliers were dubbed 'The Ring', a term that had sinister connotations. It tended to be used of the five suppliers of telephone exchanges, but there were six other agreements covering telephone apparatus, cable, batteries, cords, transmission equipment and loading coils. To the Post Office this method of doing business had definite advantages and was not seen as anti-social. It simplified transactions in that negotiations were with each industry as a whole and not with a number of separate firms. Ordering was quicker than by formal tendering and the programme for the year ahead was known. Economies of scale were achieved in the production of a range of standard items. Costs for each numbered item were determined and formed a comprehensive schedule to the formal agreement setting out all the terms and conditions, to which the seal of the parties involved was fixed. Post Office costs-investigators lived on the manufacturers' premises to monitor movements. Throughout the 1930s the trend of prices was downwards.

One of the companies' justification of agreements with competitors was the maintenance of price levels. STC was much happier to sit down with its main international competitor in transmission business in a neutral place like The Hague and sort out who was going to do what. It was much better than cutting one another's throats. On radio there was a three-way agreement: STC/Marconi/Telefunken. Originally there was a technical basis to the division of work between STC and Marconi, with Marconi making transmitters for both voice and telegraph and STC being restricted to voice only. Just as the aggressive Colonel Behn had upset Siemens & Halske by competing in Germany, the subject of a High Court action against STC in 1930, so in the same year he came into conflict with Marconi. By founding International Marine Radio Company he was competing directly with Marconi International Marine in a business that was mainly wireless telegraphy. Some technical fudging was done and it was agreed that STC could make marine transmitters for telegraphy with nominal telephony. For many years it was not a serious threat to Marconi dominance of that business.

In the dispute with Siemens & Halske STC lost the legal action but won the arbitration. The result was a draw and in the economic circumstances the two companies had to live together. Gentlemen might fall out but they had their livelihoods to consider. They felt that they were acting in the best interests of themselves and the people they employed. Their agreements, it was claimed, did not result in excessive prices to the customers. Rather they assured work and kept people in employment, which was surely a laudable motive. When a small group of men sat down to determine the price of cable or a piece of equipment they were acting in the interests of thousands.

In another sense STC was also mindful of thousands of its employees menaced by the Germans, as Richard Dimbleby told the nation during the nine o'clock news on 7 July 1939:

> At last a big industrial plant has come out into the open with its plans for self-defence in time of war. The works at Southgate cover 40 acres of ground, with a floor space of 15. There are 3,500 men workers and 2,500 women, every one of whom, except the ones earmarked for special duty, has some place of his or her own to go to when an air raid warning comes in.
>
> One salient fact emerged from today's demonstration, and was duly noted by the experts with the Lord Privy Seal. You can send an air raid warning to a factory of this size, stop all its machinery, marshal its 5,500 men and women, march them to complete safety and turn out all your special squads in exactly seven and a half minutes from the word 'go'. They did that today and, when it was over, got everyone back to work in a minute longer.

After detailing the procedures he described the conditions:

> The shelter tunnels which have been dug under a part genuine, part built-up hill at one end of the works are among the most elaborate in the country. They have sanitary facilities and fresh drinking water, and first-aid dugouts. They are proof against even direct hits, and there are benches to accommodate every one of the works' employees. The 5,500 were down below for an hour today, calmly reading, talking, knitting, or playing cards, and the temperature did not rise above 66 degrees without air-conditioning. The tunnels, which are seven feet in diameter, have twelve entrances, and three sloping emergency exits as well. They have three alternative systems of lighting, and telephones linking them with the world outside. They are in fact as safe as man can make them. So at last we have proof that a big plant can be cleared of its employees and can go into action at battleship speed. I think the moral of my report tonight should be: 'Other works, please copy'.

A Territorial Army unit had also been formed on the site. When he was visiting, Colonel Behn used to inspect the troops on parade and then have dinner with them. Shelters and soldiers would soon be needed in a less cosy, less gentlemanly world.

10

Evacuation and Growth: 1939–1945

One weekend in July 1939, just after the Southgate shelters had been shown to the Lord Privy Seal Sir John Anderson, Eric Bivand was walking in Hyde Park with a young German engineer from an associate company in Nuremberg. The German asked why people were digging trenches. When told that there might be a war with Germany he dismissed the idea as ridiculous and professed to know nothing about the Nazis. Through his visits and the purchase of German equipment the company was able to establish basic manufacture of the selenium rectifiers it had hitherto imported from Nuremberg. These were a much more efficient source of direct current since they were able to work in higher ambient temperatures than the copper oxide type, and when war came were ordered in thousands for use in power supplies in army radios, battery chargers, aircraft engine starters and many other equipments. So great was the demand that, working with a man from the Post Office, Bivand had to allocate what could be made. Output of rectifier discs expanded from half a million in 1939 to 23 million in 1944.

One thing the Germans did not supply was the formula for the mysterious ingredient 319. When supplies were cut off after 3 September its content was ascertained from a French associate company. From mid-1939 the Paris labs had the services of six or seven British engineers, among them Reeves, whose skills were not required for military work in their own country. Reluctantly but fortunately they escaped from France in June 1940, one of them abandoning his Rolls Royce on the quay at Bordeaux. At home there was plenty for them to do in what was going to be a conflict much more scientific and technical than the First World War. STC, cut off from its European associates and its contacts with New York perforce lessened, was thrown back on its own resources, resources that had to be expanded for the war effort.

At Foots Cray, where among other things cathode ray tubes, Spitfire parts, ammunition boxes, instruments and mine-detectors were made, extra buildings were put up, mainly for the manufacture of small valves. Gyroscopic gun sights used in Spitfires and Hurricanes were also produced on the site and Kolster-Brandes, which suspended radio production on the outbreak of war, set up a supplementary plant for them in the Earls Court exhibition buildings in 1943. At Southgate between 1940 and 1942 as much manufacturing space again was added as in the whole of the previous decade. That was by no means sufficient to meet the expanded demand for products and components and more space had to be found. The small but growing rectifier section moved its manufacturing to a cabinet-maker's in East Barnet, established a duplicate plant in Boreham Wood, and used a former eiderdown factory and part of a bottle factory near Southgate, sites that were within reasonable operating distance of one another. Altogether an extra 165,000 square feet of manufacturing space was acquired, including garages, a printing works and part of a sweet factory.

Like the company's other plants though, they were largely concentrated in the London area, the vulnerability of which was vividly demonstrated in the 1940–1 blitz. Arrangements had to be made so that production of important items would not be completely knocked out. When there was a serious threat of invasion all the drawings for the company's major product line, Strowger telephone exchanges, were copied and shipped to the Bahamas so that ITT had the information to carry on with manufacture if necessary. Such long-distance emergency arrangements were not generally practicable because production had to be not only increased but assured. The answer was dispersal to 'shadow' factories away from London.

From the summer of 1939 the first was established at Treforest in midGlamorgan for the manufacture of quartz crystals, important for frequency control in communications equipments like those used by Fighter Command in the Battle of Britain. The army had first shown interest prewar in quartz for tuning radio equipment vibrating in tanks and the technology, studied in the Paris labs, had been transferred to the UK and demonstrated on Woolwich Common. Further information was obtained on visits to Western Electric and its US competitors. The Treforest plant was seeded from North Woolwich, where a bonus scheme was introduced at the beginning of the war to increase productivity. Equipment and the nucleus of skilled labour were transferred from Woolwich, local unemployed miners trained in the new skills, and by Christmas the plant was in production with a similar bonus scheme. Wages increased and the company was soon accepted as a welcome change from the 1930s. The Air Ministry commended it on its increase in output. Having helped set up the plant in comparative safety, Harold Lyne reckoned he was the first man in the company to have his home bombed. A bomb that might have fallen on Cardiff Docks made a hole that would have taken a single-decker bus in the garden of his temporary flat in Penarth. Others had smaller domestic problems to contend with in settling into a new environment.

Woolwich, in London's dockland, classified as a vulnerable point, was also the site at which another vital component was made, high-power valves. During the Battle of Britain the manufacture of these was transferred with magnetic materials to a rope-works at Ilminster in Somerset, which seemed last to have prospered in the manufacture of rifle pull-throughs for the Boer War. Its floors were earthen and the machinery driven by a water-mill. The labs enjoyed the superior accommodation of a new but unused local school. There with the aid of some university boffins, many of them left-wing and glad to go all-out in support of their new ally the Soviet Union, among other things new radar valves were developed.

Initially the people of the small town of Ilminster were suspicious, even resentful of the newcomers, but the attitude soon changed as they appreciated the jobs and money that had been brought in. It was not to be so in the biggest of the dispersals from Woolwich, the move of the transmission division to Leicester, where textile factories and a dyeworks were taken over for the manufacture of line equipment, mainly for the army. Even though the trade unions were involved in negotiations on increased productivity, local labour had deep-seated traditions and did not readily adapt to a new organisation and new processes. There were strikes and the troublesome plants were a long time settling down. It was impossible to transfer the East End mateyness of Woolwich to the East Midlands. Those who were Woolwich born and bred looked forward to the end of the war when they hoped on VE Day Plus One to be back in their devastated homeland. The smallest of the four Woolwich shadow factories was a cable plant at Lydbrook in Gloucestershire employing about thirty people.

For the first time the company was operating outside the London area and its lines of communication were longer and more difficult. Co-ordinating the work of scattered plants at a time when so many jobs had a priority tag and travel was difficult turned people to the telephone. Where they would have thought perhaps more than twice in peacetime before picking up the instrument it now became a natural habit. It was not there just to receive calls; it could be used to get things moving. Costs of trunk calls were a secondary consideration when there was a war to be won.

On the home front one of the unspectacular jobs the company was involved in to keep communications going was the installation of by-pass cables and carrier equipment around major towns that were focal points on the network and also likely to be bombed. Sites of particular importance in the trunk network were the Citadel, the duplicate exchange in the City, for which transmission equipment was supplied, and the repeater station in the Whitehall Tunnel. Similarly, three-channel carrier and other improvements were introduced to the Great Western railway, before the war primarily a passenger network running on timetables but now called upon to intersperse much more freight. Handling these unexpected volumes of new traffic, troop trains and armaments, could not be done with the odd telegraph key. It needed a telephone network. The government, not a spendthrift on the Post Office in peacetime, now appreciated that communications were essential to the maintenance of as orderly and efficient a life as possible, whatever the emergency or damage. In the war, during which civilians were involved on an unprecedented scale, the telephone came of age, like that younger medium the wireless.

For the war of words the company supplied the BBC with transmitters. The first was a medium-wave 140-kilowatt unit originally intended for Lithuania that was installed at Brookmans Park to increase the power of the European Service transmissions. Mobile transmitting stations, each housed in two motor coaches, were hurriedly built for use by the BBC in the event of invasion after Dunkirk. Altogether fifty-five broadcasters, with powers from 2 to 130 kilowatts, were supplied for use in psychological warfare, the entertainment of troops, and the rehabilitation of liberated countries. Places to which they were shipped included Belgium, China, India, Iran, Malta, the Netherlands, and South Africa. In 1943 the most powerful short-wave radio station in the world was built for the BBC and installed at Skelton near Carlisle, with Marconi and STC each supplying six transmitters. Short wave could reach countries out of effective long- and medium-wave range and was also less susceptible to jamming. Communications transmitters for point-to-point and ground-to-air service were made on a much larger scale, the numbers amounting to 2,500. An unusual application of a 5-kilowatt low-frequency amplifier was at the National Physical Laboratory for making accelerated fatigue tests on air frames.

The bulk of the company's effort was directed towards military communications. Since before the war it had been working on the Defence Telecommunications Network, the acronym DTN standing to those who knew for Don't Tell Nobody and in reality amounting to a telephone and teleprinter network that linked all home defences, especially Fighter and Anti-Aircraft Commands, ensuring communications with observation posts, airfields, gun and searchlight centres. Rapid relaying of intelligence was vital for early warning and effective counter attack. It was an application involving thousands of miles of cables, switchboards and the voice-frequency telegraph equipment that had been so successful in Post Office long-distance links since 1932. The company had also been doing work on message scrambling.

Now there were all kinds of urgent special jobs. Amy Baker was one of the housewives who went into Woolwich to do her bit, initially on a month's trial at £2 a week:

> Not long after joining we went on to night work on this Navy job for 14 nights right through, eight till eight. We used to throw cherry stones at those who fell asleep in the small hours. Some couldn't keep it up and fainted. When we'd completed the job it was rushed off to Portsmouth.

As far as they knew their 'hush hush' work was for submarine detection. Another of the urgent jobs of mid-1940 was a six-repeater installation for Narvik, the iron ore port in northern Norway. L. A. Martin worked on it:

> The equipments were entirely new in design, involving new panels and bays. Nevertheless from the word go the equipment was engineered, manufactured and shipped within one month. Unfortunately the British evacuated Narvik (10 June) before the equipment arrived.

Nearer home, to provide emergency communications for the British Expeditionary Force retreating to the Channel coast, equipment was assembled and tested in a forty-eight hour stretch and flown from Hendon. At home a new network reporting system had to be developed so that information on enemy aircraft movements received by telephone in the underground centre of Fighter Command at Stanmore, Middlesex, from the coastal radar stations could be distributed more quickly to civilian defence authorities.

In the exigencies of war developments were speeded up. 'Doc' Field, a chemist, had been working in the labs at Woolwich on the development of rubber cable:

> We had never made rubber cable at all before in the company and the first was degaussing cable at the beginning of the war to wrap around ships, making them non-magnetic against magnetic mines. We used to buy in the rubber compound, extrude it on an old machine, and then vulcanize it. One of the biggest problems was that the compound had bits of cigarette packets and things in it. You don't make very good cables like that. We really didn't know the first thing about making rubber cable and we found out every bit of it the hard way. Ultimately we produced quite a good lot of degaussing cable.
>
> But we realized that we would never be successful unless we produced our own rubber compounds. There wasn't enough room in Woolwich to do this so we took on a small factory at Enfield. About this time somebody had discovered that Western Electric were operating a continuous vulcanizing machine, which extruded the rubber compound round the conductor and vulcanized all in one operation. One of these machines was ordered from America but when it was delivered they discovered that it needed a high pressure steam supply and something like 120 feet of vulcanizing tube. They didn't have that length of space at Woolwich so they put the machine at Enfield along with the compounding. That's how we started in the rubber cable business at Enfield.
>
> Early in 1942 we lost Malaya and all supplies of rubber. So we had to use other materials. That's when we started getting on to plastics, thermoplastic materials like PVC and polythene. The plant at Woolwich was quite unsuited for using these new materials and we managed to adapt the Enfield extruder to take them. With thermoplastics you don't vulcanize and instead we had long cooling troughs. Then we put filters in to keep the materials clean. With very thin insulations you must not have a particle of dirt of any sort that is bigger than about half the radial thickness.
>
> We were working for all three Services, producing tele-

phone, radar, all sorts of cables on this one machine. The output we got from it was absolutely incredible. We started this factory up with untrained personnel, people like milk roundsmen and lorry drivers. We had to train them from scratch and the people who were running the plant didn't know anything much about it either. We used to sit and argue it out. We were in great difficulty with only one machine. One of the features of the continuous vulcanizing machine was that it had a very restricted gear range and we had to extend that for plastics by putting in two old Vauxhall gear boxes we got from a breaker's yard. Doing things like that we created a new business for the company.

Transmission equipment that had been designed and manufactured for permanent installation had to be adapted for field use by development engineers like Alfred Delamere:

We started off the war with equipment mounted in bays for putting in wooden boxes. You had a nine or six foot rack inside the wooden box that had to be carted around. It was just impossible. We had to take that equipment and break it down, make it transportable in smaller boxes. Not only did we help the ministries develop systems, we had to develop many of the components that went into them. Manufacturing in transmission and design in transmission and components were all working hand in glove. The war brought us together. It was a change from the attitude of the engineers sitting back and saying 'We designed this. You make it.'

Space and weight saving were also important in aircraft radio, the main wartime occupation of the radio division. The high-frequency set developed prewar as a private venture for fighters became an effective anti-submarine weapon in aircraft escorting convoys and was officially christened the Battle of the Atlantic set. Much of the work involved the Royal Aircraft Establishment at Farnborough and the company's main customers were the Fleet Air Arm and Royal Air Force. Nearly 2,500 wireless sets were produced for use in tanks and armoured vehicles, notably for the Eighth Army in the North African campaign but, not including navigation receivers, 31,000 airborne equipments of eight different types, six of their own design, were made. Volume units were assembled and wired on a rubber belt line. In addition, fourteen daughter firms working to the designs were supervised by the company and together exceeded its own output.

One of the items was a rescue set basically copied from a

German unit picked up out of the sea and used by crashed air crew in rubber dinghies. Lifeboat radios were produced too by International Marine Radio, which as well as making communications equipment went into the business of self-priming hand pumps. Initially it moved its head office from Connaught House in Aldwych to a large house three miles from Biggin Hill, Kent, one of the main fighter stations of the RAF, and after the disruption of the Battle of Britain conducted most of its business from Southport, Lancashire. Perhaps the most intriguing unit to come from STC was the VHF homing receiver developed in a fortnight at the request of a cloak-and-dagger body that called itself Interservices Research Bureau. The equipment enabled aircraft to home on a very low-power transmitter placed by members of the resistance in occupied territories so that supplies could be dropped accurately or targets pinpointed.

Work on communications and navigation equipment was to be profitable also in the postwar world with the expansion of civil aviation. In navigation systems, what had been the German-designed Lorenz beam approach, licensed before the war, was redeveloped and given the new name of Standard Beam Approach. As it became the standard approach aid of Bomber Command throughout the war the Germans got some of their own back. Altogether the company made 385 ground beacons for sending out signals and some 36,000 airborne receivers that indicated whether an aircraft was on course or to the left or right of it. The Germans developed the technique to produce long-range directional X-beams that could be used as accurate flight paths to targets like Coventry.

A battle of the beams ensued between STC and its German associate, Lorenz. At the request of the RAF, STC developed transmitters to deflect the German beams and to interfere with their signals. The company also supplied information on the importance of the selenium rectifier, developed by its associate company in Nuremberg, to the Ministry of Economic Warfare, which was assessing the importance of German industrial targets for Bomber Command. Presumably information was supplied on the Lorenz plant as well. Whatever damage was done though was partly compensated for in 1967 when the parent company ITT was awarded $27 million by the Foreign Claims Settlement Commission for damage to its German properties.

After initially being excluded from radar business, STC made 1,400 equipments for anti-aircraft gun laying and long-range early warning. It also made 1,200 radar interrogators known as IFF, Interrogator Friend or Foe. The most urgent of the radio counter measures (RCM) was the last, just after D-Day and carrying absolute priority from Churchill himself. It was code-named the Z job on the excuse that the company had had more than its fair share of RCM panics and if only it could handle this one it would not be bothered again. Those who knew what it was about felt privately that if they did not handle it successfully they would not be around to be bothered anyway.

At the time the V1 flying bombs, 'doodlebugs', were landing on south-east England and, unknown to the public, the V2 rocket, which might carry an atomic warhead, was to follow. To direct their rockets from their Dutch launching sites the Germans were using a radio beam technique. Following this beam for the first twenty miles of its trajectory, a rocket was in effect being shot up a twenty-mile-long gun barrel. Beams were switched on only a minute or two before firing and to escape jamming a different frequency was used for each shot. STC's task was to produce a 150-kilowatt jamming transmitter capable of being tuned almost instantaneously to any frequency notified by patrolling aircraft. A broadcaster was commandeered from a BBC station and other equipment manufactured for installation on the Suffolk coast within the twelve-week deadline. Abruptly the job was stopped. The Germans had decided to change from a radio to an inertial guidance system.

A top-secret project the company did complete was for the Admiralty. A number of special low-frequency transmitters with an output of 50 kilowatts were designed and built as mobile installations. Each installation consisted of two vehicles and a pair of 95-foot sectional masts. Two stations were to be driven in synchrony from a central point. Later the company learned that the scheme was for the guidance of invasion craft to the Normandy beaches on D-Day, 6 June 1944. In preparation for the landings the company had to increase the amount of its regular telecommunications work to provide the necessary telephone lines and equipment for the railways and other services. For the Allied invasion headquarters on the western outskirts of London it provided in ten weeks a twenty-position switchboard with 160 private and exchange lines and 600 extensions. As D-Day approached, the board was enlarged to thirty positions, 260 exchange lines and 700 extensions.

Terminal equipment was made for cross-Channel cable communications to the Allied forces, a telephone system for the Mulberry Harbour prefabricated dock installed off the Normandy coast for the unloading of supplies, and lead sheathing as part of PLUTO, the undersea pipeline for carrying fuel from the UK to Europe. When this project was being mooted army experts seriously considered blowing down a recently completed four-storey building at Woolwich so that it would not obstruct direct loading on to a barge moored at the wharf. Spencer and Pheazey reacted unfavourably to the idea.

As well as its collective achievement the company made a contribution through individuals. Revealed after the war as the most notable was the work of Alec Reeves at the Telecommunications Research Establishment (TRE), Malvern, on OBOE (Objective Bombing Of Enemy). Since escaping from France Reeves had overcome his objections to working on offensive weapons, after some heart-searching justifying his decision with the argument that by the application of scientific method fewer lives would be lost. OBOE, the blind bombing and marking system controlled from the ground, was designed to give bomb strike accuracy of better than thirty yards, 'the most precise bombing system of the war'. From two ground points the range and hence the position of a bomber were determined and fed into a computer along with the position of the target, aircraft speed and direction, height, wind velocity and bomb weight. The home-based computer issued the bomb release instruction to the bomber over German territory. Experimental equipments were built at TRE and production models at Southgate, which put them into service on the South and East Coasts. For the project it had to develop an electrical timing device accurate to 1/1,000 of a second, an important precision item in destroying from above the clouds the launch ramps of the V1 rockets and in the Normandy landings distinguishing between Allied forces and German coastal batteries and defences. Reeves's work earned him the OBE.

Leo Firnberg, who had a good knowledge of Europe from his prewar work on cable installations, worked at the Government Code and Cypher School at Bletchley Park on direction-finding, locating the position of call signals. Other engineers were seconded to various government departments. From March 1942 Kipping, whose Southgate factory had become a controlled establishment under the Ministry of Supply in June 1940, applied his industrial experience on a wider scale as head of the regional division of the Ministry of Production, where his job was to remove inefficiencies and raise output. Behind him he left Pheazey as works director for the company and a tradition of improving productivity. For example, in the production of rectifier discs the original German process had been mechanised so that only one-fifth of the labour was now required. Heat treatment techniques developed in the manufacture of communications equipment were applied to the production of artificial eyes. Other developments saved scarce raw materials. With the imperative to survive, much was achieved in a short time.

Less tied to its transatlantic apron strings, the company came to maturity. Early in the war instead of a small legal practice in the Temple it began to use the services of City solicitors Slaughter and May, initially for securing quick property leases. The move had been contemplated for some time because the company felt it needed a broader range of legal advice but it took the war to precipitate the change. Distanced from New York, Spencer was able to strengthen his hand as managing director, taking decisions on the spot as a man whose prime responsibility was to the British government. Two outside English directors were appointed to what had hitherto been a tame internal board. One of them was Colonel Charles Ponsonby, the Conservative MP for Sevenoaks and Parliamentary Private Secretary to the Foreign Secretary, Anthony Eden; the other Admiral Sir Aubrey Clare Hugh Smith, the best-sounding name that had ever been on the board. For the first time the company was seeking some permanent influence outside its own circle.

Spencer was the moving spirit behind the formation in 1943 of the Telecommunication Engineering and Manufacturing Association (TEMA), of which he became the first chairman. It was formed partly in response to the government, which found it much more convenient to deal with trade associations than with a multiplicity of competing companies, especially on matters like the supply of materials. An industry or sector could sort out its problems in its own forum. Partly also TEMA was a defensive organisation, a move by a number of manufacturers to protect their interests in what looked like being a very different postwar world. Firms that were of no great account before the war had prospered with government contracts and bade fair to offer serious competition. Similarly, STC was in 1944 a founder member of the Radio Communication and Electronic Engineering Association.

By then it was only a matter of time before Allied victory and, with his eyes on the postwar world, Behn soon made his presence felt once more. In August 1944 he was in England *en route* to France and the liberation of Paris at the end of the month. He rode into the city with the US forces while sporadic fighting was still going on in some parts. Francis McLean, a former STC employee and now chief engineer of the psychological warfare unit at Supreme Headquarters Allied Expeditionary Force, had come into Paris on the Friday it was liberated and when he went round to the labs on the Saturday morning was surprised to see Colonel Behn already there. Two days later Behn was host to about a dozen people for a lunch of oysters, sole, steak and good wine in a small restaurant behind the Opéra. McLean, the only non-ITT guest, was invited for old times' sake. Behn's presence attracted attention and the British press voiced the criticisms of its own businessmen aggrieved that unlike the Americans they had not been able to get in to take care of their interests.

Behn soon got things moving again. For example, not publicised was the visit of two Englishmen, Leslie Long and Marcus Thornton, to officially neutral Spain in November and December 1944. They flew out from Hurn, the airport near Bournemouth, after midnight in an unheated Dakota, keeping warm with blankets. Their ordeal was rewarded by a stay in the Ritz and the pleasure of shouting down Germans in restaurants in English. Their mission was to plan improvements to the Spanish network with twelve-channel carrier systems and they enjoyed doing it before returning home in time for Christmas on a plane with Sir Samuel Hoare, ambassador to Spain on special mission. It would mean postwar business for the company, which had passed the peak of its wartime achievements in production. The period of greatest effort was in the middle years of the war, in 1942–3. At Southgate the labour force, which had risen to 10,500 by late 1943, gradually decreased to 9,000 on VE Day. Clearly there would be no or a much reduced demand for many of the products in peacetime and, while Long and Thornton were in Spain, after meetings Spencer had had with his division heads on postwar strategy, the publicity department was sitting down with the commercial people to assess the anticipated promotions they would be involved in.

Like other technically-based companies that had survived the 1930s, STC was bound to grow in wartime. Especially to those who had known the inhibited developments and leisurely progress of peace, the number and pace of changes were unbelievable. What else might have been achieved in the 1930s had the necessity and the resources been there? Many people who had never dreamed of working or working again were needed by industry. At Southgate the 60/40 male/female ratio of employees was reversed in an enlarged labour force. In 1942 about 1,000 women were taken on for part-time work of an average of four and half hours a day. Still more labour was required so homeworkers and outworkers were employed. Homeworkers did light assembly and inspection jobs for at least twenty-five hours a week. Outworking units were established in firms on non-essential work by using their labour and premises and supplying them with equipment and expertise. Thus shop premises were equipped with work benches or turned into packing rooms and BBC staff came in two or three evenings a week after their normal jobs to assemble plug and lamp sockets for telephone systems to be used by the Services. Many girls, most of them without previous factory experience, were brought over from Ireland for this kind of work. During the later years of the war more juniors, between the ages of 14 and 16, were engaged.

Bringing in new people and supervising them outside the company meant an increase in training, not only in job skills but in the ways of industry. Some thought that they had been misdirected in that they were not working on guns, shells, tanks, aeroplanes or other war weapons. Lectures, films and posters were used to put over the message that without effective communications sons, sweethearts and brothers in the field would be helpless. Many recruits came with preconceived notions about the drudgery and monotony of factory conditions. They had to go through induction courses before they could become useful and productive employees. Because the company had expanded and dispersed largely without calling on outside aid there were many internal promotions, with people being called upon to do more and bigger jobs. Supervision from the works manager downwards had to acquire new skills; many had to learn how to teach so that they could instruct others. Among the first to be trained were planning engineers, who received intensive training from an inspector of technical education who had been seconded to the Ministry of Labour and National Service and had studied methods in the USA. Techniques of quality control originating in the USA were also introduced.

From 1941 management and shop floor regularly sat down

in equal representation on Joint Production Committees to identify and remove obstacles to output. Within factory boundaries, now strengthened by barbed wire and day and night security patrols, people were less divided. Under the camouflage, in various ways they were all in it together, united by the long-term need to win the war and by daily common dangers. Having evacuated her daughter to the Hampshire countryside, Amy Baker was sticking it out at Woolwich:

> Early on, as soon as the siren went we used to go to the shelter. We spent a fortnight in the shelter and there was no work done. The management announced that anybody who wanted to work when the siren had gone could. Everybody did.

Shelters were not the sweetest smelling of places but that was not the reason for staying above ground. Work areas too could get hot and stuffy when the blackout shutters were put up. People wanted to get on with the job and drew courage from the presence of others. Away from the factory, with her husband on police duty, Amy could be lonely: 'I used to hate the weekends, especially when there were a lot of raids.' At lunchtime there were entertainments in the canteen, occasionally somebody of the calibre of the uniformed Richard Dimbleby talking about his experiences as a war correspondent, a comedian like Arthur English, or Pouishnoff playing popular classics at the grand piano, but some of the ENSA acts were so appalling that they were a laugh for the rest of the afternoon.

During the day there were spotters on the roofs to give warning of any incident, with instructions not to sound panicky, simply to say 'Take cover' twice in a firm voice. Information that the other defence services were getting was relayed to them over a loudspeaker circuit and they could follow on a map where the planes, or later the doodlebugs, were heading. At Woolwich the company's public address system was used to relay warnings to local inhabitants. On one occasion the telephone line from the spotters on the roof got inextricably linked with the public address system and the whole factory heard: 'There's one coming now. Yes, it looks like a mine. You'd better give them a take cover. Yes, take cover, take cover. Hold it, hold it. Thank God, it's on Siemens.' At night there were fire-watching and other security duties. The Home Guard unit at Southgate had a

Bofors anti-aircraft gun mounted on a lorry. It was never fired. As well as the menace of bombs, mines and incendiaries, early in the war there was the possibility of the enemy literally at the gates and instructions for rendering factory services useless were kept under lock and key in a steel container in the air raid precautions control centre on site.

More cheerfully, there was voluntary work to do in aid of the forces. Money for the Merchant Navy was raised through collecting boxes for 'ship' halfpennies, while dances and other functions provided funds for other Services. Les Buxton, who had joined the company only in 1939, throughout the war had about £1 a month sent to him wherever he was, even in Ethiopia. In their spare time girls knitted for the forces, three of them managing to make 600 pairs of socks. Workers in Britain were not forgotten by their associates overseas, who had never seen them. STC Australia sent to Woolwich food parcels, which were split up and presented to individuals recommended by foremen for their part in the war effort. Visitors from the Services, ministries and overseas, including in 1943 some Russian guerrilla leaders, came to see for themselves how things were going at the bench.

Churchill visited the Woolwich plant in September 1940 thanks to the *chutzpah* of the works ARP officer, Sam Catterson. After a big raid on the docks, to raise morale Churchill went to see the extent of the damage for himself. To STC employees making their way to work it seemed that there was devastation for miles around and that their own plant had miraculously been spared like St Paul's amid the smoke. Wanting Churchill to see something that had survived and was carrying on, Pheazey, the works manager, deputed Sam Catterson to make an on-the-spot invitation. Looking distinguished in his black homburg, Catterson barged through the surrounding aides de camp and despite their protests got what he wanted from the great man himself. The unscheduled tour conducted by Sam, of necessity quick, was a great tonic to the employees but a personal disappointment to Pheazey, who did not meet Churchill.

Being near the docks, the Arsenal and a concentration of factories, Woolwich was the most vulnerable of the company's plants. In the great raid of 7 September 1940 high explosives rent the tanks of Loders & Nucoline containing normally solid coconut oil, which melted in the incendiary bomb fires, flowed into STC's drainage system and blocked it for several weeks. In addition, water from the Thames fire floats was poured on the fire in such quantities that it spread through the subsoil and flooded the underground shelters, rendering them useless for a time. Although the Luftwaffe returned to stoke the fires, actual damage to the company's property was confined to a few one-storey buildings, thanks to those who pushed incendiaries into the river and the efforts of the works fire brigade. However, electricity, gas and drinking water were cut off and it was not possible to run the canteen. From the other side of London and fifteen miles away as the crow flies, Southgate came to the rescue, preparing cold meals and filling tanks with drinking water. When these could not be delivered on routes blocked by unexploded bombs, Southgate arranged for the midday meal to be taken by police launches from Westminster to Woolwich, where it arrived only one hour late.

That same month when a landmine hit Henley's canteen next door the blast knocked out many of STC's windows. In April 1941 a building that was only four years old was struck by two high-explosive bombs, which damaged the steel structure and the workshops. Damage was not caused again until the V1 raids of 1944. In July a doodlebug gliding down into STC was deflected by its right wing striking the flagstaff, and making a U-turn – the spotter said 'Take cover' three times instead of twice and was later reprimanded – fell into Henley's premises. An STC building collapsed and another was severely damaged. In clearing up, messages had to be run by hand because the relays in the company exchange had been shaken out and the telephones would not work.

On 30 June just after lunch the company secretary, John Buckland, was getting ready for a board meeting at Connaught House:

> There'd been a series of black warnings all the morning and I cussed the whole staff like anything – I couldn't get anything done. They were all downstairs. I went to the boardroom with the papers and spread them around the table and the next thing I knew the whole lot had gone up, me with it. This blessed V1 had come down in Aldwych. I remember the table collapsing on me, the place going on fire, getting up and standing on something that went through my foot – glass – which woke me up, finding my way out of the boardroom, straight out not through the door – there wasn't a wall – and coming out on the top of the stairs. But there were no stairs, so I turned round and went down the back stairs. The next thing I

> knew I was fighting with the carpenter. I was covered in stuff. I remember seeing some armbands all around an air marshal's arm and I was rescued by an air chief marshal. They were next door and I owe my life to the fact that I was mistaken for an air ministry official and was carted off to the Middlesex Hospital instead of queueing up for attention at Charing Cross, where so many people I knew were waiting until late evening before they had wounds dealt with.

Six of the staff lost their lives. The minutes of the previous meeting were bloodied and the attendance book for 30 June laconically recorded: 'No meeting held'. While Connaught House was being renovated some staff were evacuated to Southgate. It was there that the worst of the casualties occurred, on 23 August, when out of a low cloud a V1 fell between two buildings as the night and morning shifts were changing. Altogether twenty-one people were killed outright and twelve of some 200 taken into hospital died later. At the peak of the attacks over 100 of these flying bombs were logged in one night by spotters at Foots Cray, which although it was in Bomb Alley sustained only one small high explosive and a few incendiaries during the whole course of the war.

For the company it was a war of achievement, greater than in the First World War. Personnel had doubled from 12,000 in 1939 to a 1943 peak of over 25,000, new sites had been acquired, new products and methods developed against unusual difficulties. Yet it was a labour-intensive achievement and, come the end of the war, the labour was tired. A Labour government had been elected to bring the New Jerusalem and it was time for the workers, who had strengthened their bargaining positions through their unions and co-operation in the war effort, to begin to sit back and take things a bit easier. Relationships with management could never be quite the same again after you had been in the shelters together, dived for cover from the same bomb, endured similar tribulations. Indeed, junior managers who believed they could create a brave, new, more egalitarian world were joining the Association of Scientific Workers. It was not going to be like the aftermath of the First World War. But, as then, the rules of the game suddenly changed. Now that the military fight was won it was a matter of adjusting to peace and fighting the economic battle once more in a very different world.

11

Latter Days of the Gentlemen: 1945–1959

Hopes for the peace were high. Sir Frank Gill, knighted in 1941, began the first postwar board meeting with a pause 'proper as a mark of thankfulness for the deliverance vouchsafed and in memory of the fallen'. The mood was that their sacrifices should not be in vain and that a better world was going to be built through continuing Allied co-operation. Publicly, practical communication between wartime buddies was promoted by the reduction in the transatlantic call rate from the 1928 charge of £9 to £3 for three minutes.

In September 1945 in New York Behn held his Victory Development Conference, bringing together his senior managers from around the world, many of whom travelled on the *Queen Mary* on her last voyage as a troopship complete with war brides and children. Rising 64, Behn was anxious as ever for some spectacular achievement and did not want to hear too much about long-range plans. One minor development that emerged was an automatic reservations system, Intelex, designed by STC for airlines and demonstrated in New York in 1948. Based on the working of existing electromechanical telephone exchanges, it had limited application on the Pennsylvania Railroad. The possibility of electronic exchanges appealed to Behn but not the prospect of waiting ten years before they became saleable products. Even that time scale was an underestimate.

STC was faced with the problem of regaining technical leadership in the products of peace. During the war, in the national interest it had to pool many of its commercial secrets, its designs, specifications and manufacturing tricks of the trade so that, if one of its plants was put out of action, competitors like Ericsson, GEC or Siemens were an alternative source. In new products there was threatening competition from smaller companies that had prospered during the war, companies like E. K. Cole, which had started with Eric Cole making eliminators over a high street shop and was now in radar. The company also had an experience different from its Continental competitors. Whereas in companies like Philips, where production had been slowed down to hinder the German war effort, there was an incentive to get moving again, the attitude in Britain was to relax and then recover. People wanted to get back to what they had known: women to the home, men to the jobs that women had done, craftsmen to their status within the shop floor hierarchy temporarily disturbed by dilutees who had been directed into war work. A vogue word and major concern was reinstatement.

By the end of 1945 the major reduction of STC's manpower had taken place, from nearly 24,000 the previous year to just over 16,000. Even so, this was a third more than prewar and their outlook was different. At the end of 1946 the workforce won a reduction of the forty-seven-hour week to forty-four and a five-day week to boot. When Harry Gill, Sir Frank's son, rejoined Connaught House in 1949 he could not help but notice:

> It was very different to what I thought it would be. I was expecting a lot better discipline because people had had it in the Services in the war. But the whole atmosphere was 'Couldn't care less'. Coming in in the morning, if you were late your pals would look at you and say 'Can't you get here on time?' But I never noticed any move on their part to stop discussing what had been on television the night before.

When the war ended break clauses on government contracts came into force and, although ministries were settling up their accounts, the company had the problem of adjusting its operations to peacetime products and levels of demand. Production had to be kept going during the phased return from some of the provincial plants. Leicester and Lydbrook were evacuated, with disturbance allowances being paid to those who moved back into what were often far from easy domestic circumstances. Ilminster and Treforest nervously stayed, Ilminster wondering when it was going to move back and Treforest, where prewar memories of unemployment were vivid, when it was going to be closed. The company was committed to the provinces, however, and, following government proposals for the redistribution of industry, would expand there. It was impossible to get everything back into prewar plants.

When Paul Delves-Broughton, who had been in at the beginning of rubber cable at Woolwich, was demobbed from the RAF, the company was glad to take him back and sent him a ticket to Newport. Looking forward to the Isle of Wight and sailing, he presented his ticket at Victoria Station and was referred to Paddington for Monmouthshire. What had attracted the company there was a very economical rent on a Royal Ordnance factory. When it started there on 11 November 1945 gun-making plant was still being removed and the small area of the shop floor on which a training school for cord-finishing operators was being set up was surrounded by batteries of machines producing commando bayonets. As armament machine tools were removed, textile-cable machines were installed and some of the transmission equipment assembly transferred from Leicester. Physically the largest move was of rubber and plastic cable equipment from Enfield, mixing, insulating and extrusion machinery that had to be established on new foundations, for which 70,000 cubic feet of soil had to be excavated.

About the only things of use in the old factory were the overhead cranes. The biggest problem though was labour. Many skilled people did not want to leave the London area, some were still in the Services, and it took time to retrain the reduced labour force of the former factory. All these problems of postwar resettlement were compounded by the organisational split between technical sales and manufacturing, with the result that for the first two years the rubber and plastic cable activity was not a success. In 1948 the dual control was ended and the whole organisation put under one man, Don Shrimpton, who began to make it prosper.

One ambitious project that came to nothing after nearly a year's planning was building a rod rolling mill and wire drawing plant that would also supply other cable-makers. A more modest investment that was intended to stay modest was made in re-establishing the research labs, closed since 1931, at Enfield under T. R. Scott. This time their work was to be more practical, initially concentrating on the development of multi-channel microwave links for the TV service revived in the same year, 1946. Another development came out of the prewar diversification, when ITT had tried to get the company involved in handling all sorts of non-traditional products like air-conditioning and ice cream machines, for which it did not have the organisation. Things it could sell, like hand tools and small bench machines for heat treatment, had been the profitable responsibility of a unit called Stanelco, which immediately after the war became the industrial supplies division and under the energetic Reg Ballard enlarged its scope by making an attack on the private communications market.

Hitherto jobs had been taken by accident without any individual organisation behind them. Ballard determined to go for the private automatic branch exchange (PABX) linked to the Post Office network as distinct from the purely internal PAX system, in which contractors like Telephone Rentals had done so well. Initially the business was done on prices agreed by all contractors and in this non-competitive situation the only way to distinguish oneself was to build up a reputation with individual customers. One organisation to grow into a good customer was the Admiralty, which among other things wanted a rugged shipboard system that could also connect with the land network when in port.

The immediate postwar years were not a period in which it was easy for companies to be enterprising. It was a time of material restrictions and government controls deemed necessary in the massive adjustment to a peacetime economy, at its lowest point in the big freeze-up of 1946–7, the coldest winter so far this century, when plants almost completely shut down for a few weeks because of the fuel shortage. With men legally guaranteed their jobs back, the number of STC employees was fairly constant at 16,000, to be supported on a shorter working week with much opportunity for improvements in productivity. Market demand was limited, with sales and profits not really picking up until 1948. In 1949 came the 30·8 per cent devaluation of the pound to $2. 80, which make it harder to provide equivalent dividends to ITT. For the company too it was the end of an era, with the past going and not too much new being put in its place. In 1946 Norman Kipping, knighted for his wartime work like two other members of the board, Tommy Spencer and Francis Brake, did not return to the company, where he believed he would succeed Spencer, but became director-general of the FBI, the forerunner of the CBI. Pease, a former managing director, died in 1947 and in his memory a readership in telecommunications was endowed at Imperial College. In 1948 the oldest man connected with the telephone industry, the now blind

J. E. Kingsbury, died in his 94th year, leaving for his lifetime's work £109,338.

Frank Gill was 84 at the time of his death in 1950. He insisted on working to the end because he was paid in dollars, a hard currency valuable to the country. Fittingly, he died in Geneva, where he had gone to attend a meeting of CCITT, the body his inspiration had founded. His son saw him off at Victoria Station:

> He hadn't even done his shoes up. He couldn't because his feet were so swollen. In the hotel in Geneva he said to my sister: 'I don't think I'll come down to dinner tonight. I'm feeling tired. I think I'll just go to sleep.' And that's exactly what he did do. My sister went down and had dinner and when she came up he'd died.

The sum he left was not very different from his friend Kingsbury's, £105,671. Spencer, 63, in 1951 succeeded him as chairman, retaining the office of managing director.

The dual responsibility was temporarily convenient. Spencer personally needed work to fill the gap left by the death of Grace, his wife for thirty-five years. He was also the lone figurehead who could inspire loyalty and hold the enlarged company together. With the departure of Kipping, the only other internal candidate was a man of much smaller stature, Pheazey, the works director, whose elevation in the opinion of most would have been a disaster. Nor was the company prepared to look outside. Its attitude and progress were an uncertain mixture of ancient and modern, like that of its major customer, with whom it had to stay in step. In common with other government departments and as telecommunications had come into their own during the war, the status of the Post Office had increased and would continue to increase. On a practical level the size of its network would grow and the biggest postwar advances were in equipment for transmitting telephone calls and television signals.

Whereas in exchanges the Post Office was tied to a single technology, in transmission it was prepared to experiment to a degree. A repeater station was a crude piece of engineering to Alfred Delamere:

> You just took two bits of angle iron, stuck on panels and wired it all up. A couple of tons to shift and very difficult to install. What could we do? The Post Office was in trouble too because it hadn't got the buildings and was considering other accom-

modation to get communications under way again. It wanted something that could be carried upstairs and used on an upper floor, of a private house if necessary. That's the way NEP was born – new equipment practice.

It was just a light metal framework that we tried to keep down to six foot high. You'd put it in position and plug the panels in. The Post Office said it couldn't rely on plugs and sockets but was prepared to take the new equipment provided there were alternative points that could be soldered across as well. An awkward and expensive design of plug and socket was produced to meet that request.

In coaxial cable systems production took up where it had left off in 1939, with 600-channel equipment. Development of higher quality polythene by chemical manufacturers led to improved characteristics in the cable, and parallel advances in the terminal equipment and repeaters, still using valves, meant that much more capacity could be accommodated on a single lay, first 960 telephone circuits – equivalent to one black and white TV channel – and by the late 1950s 2,700. A Post Office 1959 report described it as 'a complete revolution in the methods of long distance speech transmission. A trunk circuit can be provided in these cables for a small fraction of what it cost when a separate pair of wires was needed for each conversation.'

Just before that was written, the twopenny telephone became the new standard when the subscriber trunk dialling (STD) tariff was introduced in December 1958 based on a call unit of 2d (less than 1p). The Post Office was still faced, however, with the high cost of connecting each subscriber by a separate pair of wires to the local exchange.

An even bigger stimulus than the telephone to the development of long-distance transmission systems was television, which was growing much more rapidly as a mass medium. Before the war ended the company was involved as a technical witness in the deliberations of the Television Committee, set up by the government to plan the restoration and development of the service. In 1949, when TV began to extend to the provinces, the company supplied to the Post Office a one-inch coaxial cable between London and Birmingham for sending programmes to the Sutton Coldfield transmitter. Although it was late for the opening of the Midlands TV service it was the first cable of its kind installed anywhere and marked a significant advance in the depressing postwar years. It was soon extended from Sutton Coldfield to

Holme Moss, the Pennine transmitting station, and the company supplied some cable for the link between London and Wenvoe transmitter in South Wales.

Another transmission technique pioneered prewar came into its own, as prophesied, with the growth of national and international television. Since the cross-Channel microwave links were established in the early 1930s the company had been working on portable microwave links. These went on field trials with the BBC in 1948 and were used in 1950 on the first TV outside broadcast, of the Boat Race. Later that year, marking the centenary of the laying of the first cross-Channel telegraph cable, they linked France and the UK in what was the forerunner of Eurovision, in which from 1954 countries with different line standards could become one viewing public. It was the Coronation of Queen Elizabeth II on 2 June 1953 that gave an enormous fillip to TV, a medium initially shunned by the old guard in charge of the arrangements but which appealed to the young monarch.

In time for the event the company, whose role would otherwise have been confined to providing a public address with new microphones within the Abbey, was engaged to supply and install the longest link in the TV transmission network, spanning the 250 miles from Manchester to Kirk o'Shotts in Scotland. The system, the first to use a travelling wave tube amplifier, was designed by the laboratories at Enfield and manufactured at Woolwich which, having won the battle with Southgate that it was a transmission rather than a radio responsibility, had to change the design so that it could be produced. No expense was spared to complete the job on time and to the required standard to please the Post Office, the BBC and Scottish viewers. It was an occasion on which a break in transmission could not be afforded.

Through an improved cross-Channel link, for which the total responsibility from London was taken on by STC because the BBC was fully stretched, the coverage was exported live to France and thence via Belgium and the Netherlands to Germany. In a souvenir booklet the company claimed:

> For the first time in British history, the television camera and microphone, backed by the latest and most extensive network of transmission systems over cable and radio networks, enabled an unprecedented number of people in Great Britain and on the Continent of Europe 'to be present' at a Coronation. In the comfort of their own homes they were

> privileged, as any Peer of the Realm, to see the ceremony in Westminster Abbey, and without personal inconvenience to watch the Coronation procession from the best vantage points along the route.

In 1956 STC supplied the first UK microwave link to carry trunk telephone traffic and in 1960 a permanent cross-Channel system for TV and telephony.

Unfortunately there was not a similar success story in the company's main product line, telephone exchanges, where progress was much slower. All sorts of electronic devices had been developed as part of the war effort and electronics were beginning to be applied to the process of computing. If it was possible to do mathematical calculations in fractions of a second by other than electromechanical methods surely the principles could be applied to switching telephone calls, especially now that there was a miniature equivalent of the valve in the transistor, invented in the world's major telecommunications research centre, Bell Labs, in 1947. The first European ones had been made at Ilminster.

Electromechanical switching was basically a nineteenth-century technology and although it had been refined it had the basic defect of anything with moving parts. Contacts wore and needed cleaning and adjustment. This was true to a lesser degree of a more advanced electromechanical system, generically called crossbar, developed by a French associate company CGCT and available to STC; but, unlike its competitors, the company was not interested. Its view was that an intermediate system was a diversion of effort and that an electronic exchange, much more reliable and faster, was not far off. In 1952 it concluded an agreement with Western Electric on semiconductors and embarked with the Post Office and the other manufacturers on a project to design and build a model electronic exchange as a research project. Two years later an electronic exchange was, as always, two years away.

In the meantime the company deferred to its major customer. The Post Office insisted that it lay down specifications for equipment, it tended to claim that all development was its inspiration and that the companies were mere suppliers whereas they, for example, wanted to adopt printed circuits much earlier than they were allowed to. Post Office delegations were respectfully met by senior management and, lower down, its day-to-day inspectors had their tables reserved in company canteens. Yet the world was changing around them.

In the emerging affluent society of the 1950s the telephone, which had primarily been a business instrument and an appurtenance of the well-to-do, was beginning to become a consumer durable for Mr and Mrs Everyman. Like the car and packaged holidays it was starting to enlarge people's horizons, to extend their range of friends and acquaintances beyond the area in which they lived. In the language of sociologists, it was 'dispersing social networks' and 'widening the psychological neighbourhood'. A practical example was the formation in 1953 of the Samaritans, a group to help would-be suicides by discussing their problems remotely but closely. People wanting telephones were almost as desperate. There was always a waiting-list and when they got an installation it had to be a party line. Unlike the consumer durables provided by private enterprise the level of telephone growth was a function of supply rather than demand. The Post Office was a government department, a monopoly held aloof from the market by the Treasury and existing in daily fear of a parliamentary question.

Under this dead hand technical changes nevertheless went on: the trend from PAXs to PABXs, the advance from 1958 of STD, the introduction of cordless switchboards and, with improvements in plastics, lighter weight telephones and headsets. The transistor made possible the development of smaller compatible components, both passive and active. Magnetic materials for inductance coils, which had been hefty lumps of iron and copper, became smaller and more efficient with more closely defined characteristics. Similarly in capacitors – formerly condensers – it was possible to pack more into a smaller bulk. It seemed that the more remote a product was from Post Office influence the more likely it was to progress, a contrast that was evident to Ken Frost at Southgate:

> Switching was certainly ageing, static product, static personnel, cosy business. On the same site you had radio and radio was young people, very active, not a cosy business but very interesting, very lively. There were the two organisations on the site but they didn't talk to each other. We even had lunch at separate times.

Whereas the efforts of the switching division were almost

entirely dedicated to one product the radio division under Charles Strong had a totally different policy:

> The large repetition business is apt to be spasmodic and we have found over the years that our more stable business lies in the engineering and manufacture of a wide range of specialised equipments required in relatively small quantities. Our stability lies in the diversity of our products . . . It is often very difficult to guess in advance what line is going to develop in a big way. There is safety in numbers and so we endeavour to provide ourselves with up-to-date systems in nearly all the main sections of the field.

Work continued for the BBC with the supply of transmitters for radio and TV. Two high-power units were supplied for the regional Home Service, Westerglen for Scotland at the end of 1949, and Washford for Wales early in 1950. From mid-1954 to give better quality reception on 'steam' radio there was a market for VHF/FM transmitters, of which STC supplied more than Marconi. Strong's former employee, Francis McLean, now deputy chief engineer of the BBC, urged him to go seriously into TV transmitters after he had successfully made a number of 500-watt units. The advice was not heeded. Strong, who had an obstinate streak and the backing of STC's technical director, A. W. Montgomery, was against going into high-power units because he believed that few would be wanted and that the money would be in low-power units for fill-in and relay stations; but he was wrong.

Development of high-power klystrons enabled the technology of the big broadcaster to overtake the low-power concept and STC's interest in broadcast equipment dwindled, to the benefit of Marconi. EMI wanted STC to do the sound side of the first two big TV transmitters for the BBC but it was not interested. Come the opening of the new BBC Television Centre in 1960, STC's contribution was limited to a dozen miles of cables for telephone, sound and video transmission and a 2,000-line PABX. Similarly, the company, which had done well in both studio and lip microphones, failed to anticipate the small personal TV mikes that came from the USA and the Continent. In 1954 it intended to go into the manufacture of TV studio equipment, from the camera onwards, but nothing came of the idea.

To Strong the future was in airborne radio equipment, for defence needs, the growing civil aviation market – 'world-beaters like the Comet and the Viscount' – and export. The market was in hundreds and thousands of units, covering both communication and radio-location equipment, and accounted for over 30 per cent of his division's sales in 1953. Strong could claim: 'It now therefore appears that the radio equipment in Britain's newest aircraft will be predominantly STC equipment.' There was always a question mark over defence requirements but the trend of civil business was encouraging. In 1956 a BOAC Stratocruiser on a scheduled London–New York flight received radio-teleprinter messages with weather and navigational reports. Moreover, the market for navigation equipment, which embraced radio altimeters, receivers for ground beacons and instrument landing systems, direction-finding equipment, and precision approach radar, would expand as the skies became more crowded. As if that were not enough to keep people busy there were point-to-point communication transmitters and receivers. A control system for airfield lighting and traffic movements was installed at London Airport.

Another market in which the company was establishing a reputation was remote control schemes in the electricity industry. In 1950 it installed at Loch Sloy the world's first miniature direct-wire system for controlling hydroelectric generators. This was followed by similar equipment for the Owen Falls project in Uganda and the Kariba Dam on the Zambezi, the largest power station in Africa. It also provided with two other UK manufacturers telephone, supervisory and telemetering equipment to the Central Electricity Generating Board for re-equipping power stations and control centres. A related market was in switching systems to take the place of people handling quantities of punched paper tape in centres relaying teleprinter messages. In 1956 the company began designing the STRAD (Signal Transmission Reception And Distribution) system, which had a wired programme to follow fixed routines. This system had a life until the early 1960s and in 1961 a demonstration was staged of data being transmitted over ordinary telephone lines, a practice destined to spread with the application of computers. The company made its own foray into that new business with the Stantec Zebra, a machine unlike others in that the operator had to stand at the console.

New products were felt to be necessary not only to keep abreast of technological developments but also to open up new markets especially overseas, where by the mid-1950s competitors were making inroads into what were regarded as traditional preserves. The war had swept away the monarchs with whom Behn had dined and the political map of Europe had been redrawn by Stalin. Before the Iron Curtain came down over Czechoslovakia's western frontier STC supplied short-wave radio equipment totalling 320 kilowatts, and several visits were made to discuss improving the country's cable networks. On the last of them, during the Communist takeover of February 1948, the STC technical adviser, Leslie Long, had the unfortunate experience of having the local ITT manager, Plocek, round his neck in tears, begging for something to be done to get him out. He was hanged as an American spy.

In 1949 Edgar Sanders, the British manager of ISEC's office in Budapest, was arrested with others on charges of espionage and sabotage, soon found guilty and sentenced to thirteen years in prison. He served three and a half years in solitary confinement, during which Anglo-Hungarian trade almost vanished. When he was pardoned in response to a petition from his wife, trade talks resumed. The same judge had presided at the trial of Cardinal Mindszenty and similar methods were used to extract confessions. Led by Behn, ITT denied the charges. For STC, far removed from the rights and wrongs of this cold war clash, the end of the story was a mild embarrassment. On his release in 1953, Sanders came to Connaught House to receive compensation from ISEC for loss of earnings.

In exports STC was a model company, not needing government exhortations to recover and develop markets. In Western Europe countries were anxious to rebuild their communications networks and they sought suppliers, whose problem was not selling but allocating what they could make. Transmission equipment was supplied to Austria, Belgium, Italy, the Netherlands, Spain and Switzerland. The Commonwealth was a source of business: a direct radio-telephone link – the world's longest – was established between the UK and New Zealand, the first carrier cable installed in India to provide trunk telephone circuits, microwave systems installed in Canada for the railways, from Malta to Gozo, in Malaya, where terrorist activity dictated the design of the buildings, and New Zealand, which like Hong Kong had five-year bulk supply agreements with British manufacturers for telephone exchanges. A specially developed rural carrier system was applied in places like the Australian outback and South Africa. In a sense the Free World was the company's market: coaxial cable and microwave for Brazil; the first main-line microwave system in Japan for telephony and television, necessitating visits by Japanese engineers, who aroused shop-floor antagonism among those who had suffered during the war; radio stations for Burma; telephone networks for Egypt and Portugal; coaxial cable for Ireland; high-frequency radio transmitters linking Madrid and New York.

These were all contracts the company was proud to announce to the press and many of them made national news. They were palmy days too for sales engineers like Ken Prior:

> In the early 1950s there was virtually no competition. It was an absolute seller's market. Anything we could make effectively we could sell. We would discuss with the administration what was required, go back to the hotel in the evening, draw up a scheme and present it to the customer with a price the next morning. Inevitably he'd take it. There was no other way he was going to get it and if he really wanted it he had to place the order there and then. The state of telecommunications in Europe at that time was catastrophic. Everything was building up after the war. All we had to do really was to make sure we put the right price on it. There was enthusiasm, no difficulty in getting orders, no difficulty in selling anything. We made profits.

Public announcements of successes obscured the fact though that around the middle of the 1950s the market was beginning to shift in favour of the buyer. European competitors like Siemens in Germany were hungry for business and determined to get it. Arthur Beck noticed the change in the demand for broadcasters: 'Italy, Spain, Sweden, Australia were making their own. An export market in FM did not develop. There were many more people in the transmitter business whereas in the old days we had only Marconi as a competitor.' Markets were not so captive. In the colonial era the influence of the Post Office extended overseas as well. As there were few consulting engineers specialising in telecommunications, Commonwealth territories without technical expertise relied largely on the 'independent' Post Office for advice. Any overseas administration linked to the UK would have to buy its equipment to British Post Office specifications, even though these were more expensive. With the wind of change blowing, some overseas customers were prepared to accept something less stringent and therefore cheaper. For their part the companies believed that they could sell more overseas if they were allowed to write the specifications, as their competitors could do.

One way in which STC succeeded in impressing its technology overseas was through licensing, particularly of coaxial cable manufacture. Coaxial cable was technically and commercially a successful product. Postwar, cable was the most profitable side of the company's business and Spencer, a cable man at heart, was glad to give it all his support. Directly responsible to him in head office, one man was employed full time on licensing, working mainly with the cable people in Woolwich, where the appropriate machinery was designed and much of it made. Licences were negotiated for specific territories, with restrictions on export, in return for a royalty on sales. Establishing techniques and standards of cable manufacture in places as far apart as the Argentine and Denmark, India and Switzerland in turn helped sales of transmission equipment, which also tended to be the staple product in Commonwealth associate companies. Similarly, there was a demand for rubber and plastic cable technology originating in Newport, which also became involved in supplying know-how and machinery and in training production engineers.

In the longer term the establishment of overseas plants was bound to have a cumulative effect on export sales. Not unnaturally, countries that saw an expanding telecommunications market and were interested in becoming self-sufficient were quick to seize opportunities. Two years after independence, the Indian government signed an agreement for a cable plant expected to meet all its requirements, with exports to Burma and Ceylon as well. Plants were set up in Australia, which also served Pacific islands. No country was more interested in self-sufficiency than South Africa under its Nationalist government. There STC became a one-third partner in a telephone cable plant. Also, in 1948 it set up the country's first assembly unit for telecommunications equipment, incorporated a company in 1956, signed a transmission bulk supply agreement in 1958 and then built a manufacturing plant near Johannesburg.

There was also competition from associate companies on the Continent, where a new energetic European spirit was emerging. In theory the ITT companies worked in harmony, co-ordinating their export activities through the European

Commercial Department in London, which advised on the most suitable company to put in a tender for a specific project. On some international projects, like the first broadband microwave link across the Channel, co-operation was obvious, but elsewhere it did not always work because the companies could assert their independence and put in a bid if they felt they had a particular edge. Especially on internationally financed projects, the European Commercial Department could find itself trying to hold the ring among jealous associates. They got their own back by arguing that such a department was unnecessary, that its location and staff under Sir Francis Brake were biased towards STC, and that its costs – especially on entertaining foreign visitors – were excessive. It did not survive into the latter part of the 1950s, the period when as far as STC was concerned it was most needed.

The company's commercial director, A. McVie, had realised that market conditions were changing and in 1954 estimated that over the next seven years export business in present lines of manufacture would fall by about 20 per cent. With existing technology being licensed or superseded it was clearly necessary to find a substantial replacement. It came in the form of undersea telephone cables, an old idea that was turned into reality by advances in materials, components and design. This new form of international communications revived in Behn's declining years his dream of the 1920s, when he was establishing ITT, and opened up renewed possibilities of international politicking on the grand scale.

As a connoisseur of power, Behn knew that in the nineteenth century telegraph cables had been strategically laid with London as the centre of a world-wide Empire having a predominance in the handling of money and news and that now owning international telephone cables would bring him prestige and profit. For its part the Post Office, the British government department that thought most about the developing technology, was concerned that America should not secure a dominant position, which would weaken its own hand in negotiating international agreements. The technology appealed to Spencer, who always regarded cable as the foundation of the company's fortunes, the sound product that financed all other developments, from high-power radio to components of passing interest like the transistor. Getting into the manufacture of highly specialised cable for international markets would be very profitable.

Since the abandonment of transatlantic cable design in the Depression, the company's efforts had been limited to carrier equipment for shorter projects like St Margaret's Bay–Calais (1933), Bass Strait (1936), and Cook Strait (1937). The cable for these projects was supplied from the UK respectively by the Telegraph Construction and Maintenance Company, Siemens Brothers, and then, because there was not enough business for the two of them, their jointly owned company Submarine Cables Limited. Development of undersea telephone cables for longer distances and higher capacities depended on recent advances in materials and technology. Polythene from ICI, a tough flexible plastic with ideal electrical characteristics, had been applied by Submarine Cables in wartime contracts, notably for secure cross-Channel communications in the wake of the invasion of Europe. As on land, higher frequency transmissions had to be amplified by repeaters, which needed waterproof casings, a power supply and reliable components – finding, raising and replacing a defective repeater on the seabed could cost £250,000. Spurred on by US competition and the demands of D-Day, the Post Office had in 1943 inserted a submerged repeater in a coaxial cable between Holyhead and the Isle of Man. It was coaxial cable, with its higher capacity, smaller diameter and lighter weight, that was to be the type for undersea telephony.

In coaxial technology, but not for undersea use, STC could claim expertise. The key to transforming the economics of undersea cables as a means of international communication, however, was in the development of suitable repeaters, where again expertise was in land use. The transmission division produced various designs, starting with an American-style flexible one not unlike a length of thick cable and moving on to a cast-iron box that was attached to a cable as if it were a clothes line. Four repeaters of this type, two in each of two cables, were supplied in 1950 for a shallow-water route between the Netherlands and Denmark to give two-way operation of thirty-six circuits. F. C. Wright spent two weeks in the Hague convincing the Dutch, the prime movers on the project, that STC could do the job:

> We were much more prepared to make repeaters than the customer was prepared to accept them . . . It was the birth of an entirely new industry . . . We made the repeaters in one of the sheds in Woolwich with open doors. No clean air, just ordinary components. The cases were built on loading coil

> case principles and bolted together in the open air. We stuck them under the sea and they went on and on and on.

This type of design was suitable either for insertion in an existing cable or for laying in new cables in small numbers but not for laying cables and repeaters as a continuous process. To meet this need a new design was developed, a torpedo-like rigid tube about eight foot long that could be spliced into the cable and that was first applied in 1954 in the UK–Norway cable, where seven units were used. They were made under clinical conditions.

In the meantime developments had been taking place across the Atlantic that would put a different order of magnitude on the business. Gordon Duddridge, who had hoped to run a reorganised telephone cable division in a new factory at Basildon, Essex, was disappointed when that project was dropped. He became interested in developments in undersea cable:

> At the end of 1953 it was obvious that the Bell system was planning for long-distance undersea cables. It would be a very sizeable business, cable plus repeaters. STC were in the lead in Europe in repeater engineering and manufacture although working with the Post Office and Siemens, which owned half of Submarine Cables Limited – then the biggest undersea cable manufacturer in the world. In the system capacity they were then talking about the value of the cable business would be something like five times the value of the repeater business on a given undersea system. So STC would be doing all the brainwork on the development of repeaters while others would be getting the benefit in very big orders with cable. That was one of the main reasons why ITT went into the undersea cable business.

Behn, glad for an excuse to fulfil his ambition of laying a transatlantic cable, treated it as an aspect of the Atlantic alliance, a necessity to NATO, and in 1954 sent Rear-Admiral Ellery Stone, president of Commercial Cable, an ITT telegraph company that would be the operating body, to secure international landing rights for 'Deep Freeze', a cable from the USA to the UK via Canada and Greenland. In spite of international politicking at Cabinet levels, the project never came off but, in readiness for it, ITT put up the money for a factory to make undersea cable at Southampton New Docks. By the end of 1956, two years after planning started, it was in production with nearly £2 million worth of machinery producing two different types of cable for short-haul routes, UK–Belgium and UK–Jersey.

Duddridge was fortunate in being one of the first of the new breed of division managers to have full responsibility embracing engineering, marketing and manufacturing for his division and from the start ran a three-shift round-the-clock manufacturing system that was more efficient than the two-shift of his longer established competitor. By this time, ninety years after the first successful transatlantic telegraph cable, the first transatlantic telephone cable, TAT-1, containing sixteen STC repeaters on the Newfoundland–Nova Scotia section, was in service. As some forecasters reckoned that its 36 circuits would be sufficient for twenty-thirty years, the radio stations were closed down. Within two months they had to be opened again to cope with the increased traffic. Stimulated too by the successful launch in 1957 of sputniks, the forerunners of space satellites, development proceeded with 120-circuit repeaters being laid to Belgium and the longest cable being laid across the North Sea, a 60-circuit link between the UK and Sweden. This was a trial run for a UK–Canada cable, CANTAT-1, the first part of the £88 million round-the-world Commonwealth cable, conceived in 1958.

Particularly during the 1950s, electronic components emerged as distinct enterprises, ceasing to be appendages of the large traditional product divisions. Instead of some components being made primarily for company equipments they began to be mass produced and sold to other customers, which necessitated an entirely different organisation from that for selling telephone exchanges and drums of cable. In 1956 a components group was formed under P. H. Spagnoletti but it took another four years for it to achieve responsibility for all its sales. Recognition of the maturity of these activities came with the key to the door of factories in new towns and development areas, where the company could take advantage of grants, low rents, and housing for employees, most of whom moved to the new areas.

Ideas of workers and executives living alongside one another did not become a reality but in their new locations the businesses prospered: from 1954 in Harlow, Essex, rectifiers, magnetic materials and quartz crystals; in Paignton, Devon, from 1957 valves out of Ilminster, and capacitors from Woolwich. Conceived during the Cold War, the Paignton plant was built with an air raid shelter and by a hill into which manufacturing could be transferred if things warmed up. Originally it was intended to be the site for two active components, valves and semiconductors, with passive components being concentrated at Harlow, but the personal rivalry between two men, the group manager Philip Spagnoletti and Chris Foulkes, prevented any such logical organisation. Valves in Devon were over 200 miles away from semiconductors at Foots Cray in Kent, where the transistor division was formed in 1957, taking over the premises of Brimar valves. When Standard Telecommunication Laboratories was founded after the war the intention was that the unit should stay small, unlike the ambitious labs that had been closed in 1931, but the gathering pace of technical development had made growth inevitable. The need for more research and development area forced STL to move in 1959 from makeshift premises at Enfield to rural peace on the edge of Harlow.

For the company it was not so much the beginning of a new era as the end of an old. Behn, 74, retired in 1956 and died the following year of a heart ailment. A requiem mass was said in St Patrick's Cathedral, New York, and he was buried in Arlington National Cemetery with full military honours. It was an opportune time for Spencer, 69, to step down as managing director of STC after completing fifty years' service. The move was also a shrewd one. Net income had been falling badly in 1956, a situation made worse with cutbacks in government spending in the autumn of 1957 by the hard-line Chancellor Peter Thorneycroft. The Post Office, which had been the utterly reliable basis of the company's business, suddenly became too many eggs in one basket and it was impossible to shift them quickly. In February 1958 the stock market bottomed out, a price war broke out in the cable industry, and there was even a recession in small businesses like marine radio.

The man who succeeded to this luckless inheritance was F. C. Wright, a 53-year-old engineer from the transmission division who had spent his working life with the company. He was faced with enormous problems and he also had to correct the management mistakes of a generation. Time was not on his side. Across the Atlantic stockholders were becoming restive about the corporation's poor performance and something would have to be done. Singled out for attention would be STC, which accounted for one-sixth of the population of ITT.

12

Weary, Stale and not so Profitable: The Later 1950s

During the 1950s, as the British people emerged from postwar austerity, the youth revolution began. A young Queen was on the throne and the retirement in mid-decade of the elder statesmen of the three major parties seemed to clear the stage for new ideas. Young people emerged as consumers and in new occupations like computer programming. If they did not read the Angry Young Men and *Look Back in Anger* they did at least identify with *Rebel without a Cause* and want to *Rock around the Clock*. They questioned accepted notions of authority and morality and voted with their feet, whether it was on the anti-bomb road to Aldermaston or deserting tea-shops for coffee bars. In so many ways they were no longer immature versions of their elders but people with a distinct culture and outlook, noisy harbingers of the age of mass leisure that would be brought about by atomic power, automation, computers, and the spin-offs of space exploration.

Young ideas were not always influential. With the obvious exception of businesses that could cash in on the new markets, industry at large ignored them. Only gradually was it coming to appreciate that a change of attitude was necessary. In the words of economists and management consultants, companies were having to shift from being production- to being marketing-oriented. In a world that was becoming more competitive it was no longer good business to try and sell what the factories could make; goods had to be produced that the customer wanted.

To this trend STC was an exception. Its problem was not the market, which was largely out of its control. Most of its business was with a monopoly customer, which generally placed orders in assured volume but which could also cut them off at Treasury whim, or dither. Because the Post Office was not a marketing organisation reacting to public demand, it could not forecast, only place orders for what it could afford. In turn its suppliers, without the discipline of the market, had no need for dynamism. With largely predictable business, managements did not have to concern themselves too much with productivity and efficiency and could live in their insulated world like latter-day Romans whose Empire was crumbling at the edges. As one senior manager, referring to the bulk supply agreements, which were largely administered by STC personnel on behalf of the industry, put it: 'We were sheltered under an umbrella of complacency.' Were not the knighthoods of Spencer and others evidence that the company had at last arrived, was part of the Establishment?

At its 75th anniversary exhibition the company noted: 'In the last 30 years STC has designed, manufactured and installed over one million lines of automatic telephone exchange equipment for the British Post Office and also for Colonial and Foreign administrations.' That was in 1958, the year in which the Common Market of six European nations formally came into being. Yet STC was looking inward to its own affairs, not trying to emulate its associate companies on the Continent. True, the companies' publicity people met to produce standardised publications under their own names for a loose-leaf European catalogue, but that was as far as it went. There was little interchange of ideas in other functions and levels because to some extent companies were competitors, as was STC *vis-à-vis* other Post Office suppliers when it came to manufacturing methods. Nor was there any modernising influence from the USA. The ageing Behn had been the only visitor of any consequence and he was just concerned with broad budgets and a cursory inspection to make sure that space was being well used. On one occasion at Southgate, to fill vacant space, a complete floor was moved from one building to another while Behn was being entertained to lunch and on the afternoon inspection he did not realise that he had already met the staff. It was a small symptom of a deeper malaise.

Dr J. D. Stephenson, a newcomer who spent a brief period as number two to F. C. Wright, later summarised STC:

> The drive had gone out of the business. The Post Office business was running at 7 per cent profit, actually a bit more. The money on Post Office contracts was made on cost saving. Telephone business was the mainstay. STC relied on the Post Office to keep it going. The transmission side was something of a mess with the problems of small batch production. The

private side was running at about 30 per cent profit. Radio was making a contribution and components scraped through.

It was the private communications division that was operating in the most competitive market. Reg Ballard, who ran the division, preferred competition to fixing the price in a smoke-filled room and then having one of the members of the association find a way to give an unofficial discount.

He was perhaps an exception in enjoying the cut and thrust of seizing opportunities and getting business. More common was the situation of having a technological superiority but not exploiting the available market. The pursuit of profit as a considered aim was not quite gentlemanly. One manager made it clear to his staff that he had no desire to expand beyond his existing premises, which were full anyway. He saw little point in trying to achieve a bigger turnover because it meant employing more staff, in themselves possibly a problem because they would not necessarily be of the same company mind. With a lot more going on, control of developments might be lost. That way, technical mistakes could be made and the company's enviable reputation put at risk.

His opinion, sincerely and firmly held in what he saw as the company's interest, he defended to some of his own staff who argued equally sincerely for a more businesslike approach. Attempts were made by senior management to move him into a job where his attitude would have been more appropriate, but he refused to budge from his entrenched position. A faithful servant, he was allowed to remain where he was. Even had he been moved the company would still have had a problem. It lacked people with business expertise and there was no obvious candidate to take his place.

The flaccid organisation was run by an ageing group of inbred engineers, men who in their time had made undeniable contributions to the progress and success of the company but who had now grown old together in virtually a closed society. Promotion was from within, not necessarily depending on qualifications. R. A. Moir, for example, was director of industrial relations, public relations and personnel but he had no qualifications in any of them. He was one of Spencer's cronies. Imitation was the greatest form of flattery and a way to hold one's position in the hierarchy. Some assistants acquired mannerisms from their bosses, which in time were picked up by their assistants.

Such attitudes, amusing or annoying to those immediately

involved, had the wider effect of discouraging innovation. Kipping, the man who brought in new ideas from the USA and created workshops at Southgate that employees took pride in and outsiders envied, was succeeded by one of his pupils, a former rate-fixer from Woolwich, Reg Helbing. It was a change in the pecking order. Instead of being overshadowed by Kipping, Pheazey, now works director, dominated Helbing, who unimaginatively consolidated the work that had been done in the 1930s. His rigid manner was not likely to encourage bright ideas from below, especially when the company thought it could grow its own managers from its apprentices.

Manufacturing too was a lower order of life in the hierarchy. Gordon Duddridge, who prewar had been dissatisfied with his prospects among the old school ties of ICI and had run two government factories during the war, found that he could not get on with the chief engineer of the telephone cable division, who argued, 'The whole trouble is you're a damned traitor, Gordon. You're working in manufacturing with a university degree. You're a traitor to the engineering cause.'

Product not production engineers were well up in the hierarchy, a status difference reflected right down the line and not easily altered. Duddridge made some changes:

> A Southgate engineer-of-manufacture – that Western Electric term was still in use – often had arguments with me for the fact that I wanted people with degrees and good educations. He didn't want any. He could make his production engineers out of people who had been handling the shop routeing cards.

A 1954 report made the point:

> Such manufacturing development as is done at present is done as a side-line by the engineer-of-manufacture. It was stressed that nowhere in the system is there any manufacturing development. The engineers of manufacture would investigate any process put up to them but someone else had to originate the process. There was no one imagining processes that might be useful and then setting about making them practical. In other words, the engineers require a group to which they can talk and which is authorized to do experimental work on manufacturing development.

More fortunate were the 'real' engineers, the people who had created the company's world-wide technical reputation, as young Peter Sothcott observed in the radio division:

Practising engineers held most of the senior positions and were very largely left to operate without interference by central management. Their status and prestige, which was considerable, rubbed off on their junior professional colleagues, who thus formed a kind of elite. Junior engineers' opinions about products and divisional strategy were listened to, even if unsought, by management – and their advice was sometimes even taken. In turn, the juniors' attitudes to the company and its customers were largely modelled on those of their seniors. It was a very old-fashioned set of attitudes, too. What seemed most important were idealistic things like the pursuit of technical excellence, good engineering, customer satisfaction, service to the community, and mundane things such as keeping to schedule, or making a profit came fairly low on the scale of priorities.

A very good example of this attitude in action, and one in which I was heavily involved, was the service we gave to the RAF in Coronation Year, and which led to my spending Coronation Day in a leaky caravan in Regent's Park. The RAF needed a high precision aircraft guidance system to control the fly past on Coronation Day with sufficient accuracy to ensure that the Queen's crown did not fall off. For this, STC developed a new system, tested it, installed it in Regent's Park and operated it both during rehearsals and on the day. The whole exercise was repeated twice during the following year, once at RAF Odiham and once in a rhubarb field near the mouth of the Southern Outfall Sewer. It took a large number of man-hours and must have cost a fair old sum. Yet, as far as I know, no money changed hands and we made very little publicity out of it. The boss thought it was our public duty to help.

Engineers dealt with customers' engineers, who talked the same language and made the decisions on what type of equipment was to be ordered. There was not a separate category of salesman with a developed commercial sense and an eye to business opportunities, an omission of which Arthur Beck in broadcasting equipment was aware:

> Valve sales were not tied up with radio division sales. Marconi and Philips would undercut us on the equipment price but tie up a long-term agreement on valve replacements. Valves in a 100-kilowatt transmitter cost about £1,000 each and there were six of them, lasting about a year. The margin on valves was around 50 per cent.

This was just one aspect of an inadequate organisation and

administration. More serious was the lack of money management. 'Doc' Field, running the rubber and plastic cable division:

> We had no idea what state we were in financially. We had no figures or anything. It was just a bit of a joke really. We just carried on and I suppose somebody knew somewhere whether we were making money or not, but we never did much.

Pheazey took a broad view of costing, relying on rudimentary systems in the belief that the amount of paperwork could easily become more expensive than the economies achieved. Dr Stephenson soon realised that 'The monthly statement of work in progress with an estimation of the quality of the work etc. was just guessed at. The monthly results depended upon the optimism/pessimism of the person doing the guessing.' Sometimes customers came to the rescue of an engineer who lacked proper business backing:

> Once I went to Sweden and on the plane discovered that in my £50,000 tender they had left out a £5,000 rectifier. That was all the profit. A friend in the administration in Sweden understood my predicament and put it right for me by starting to delete things he didn't want.

In the hierarchy decisions were made and handed down, often without consultation, a practice that hurt 'Doc' Field:

> Like the decision to come to Newport and shut Enfield down. I felt it was never discussed or argued. Somebody, in this case John Pheazey, took the decision. The result was of course that about two people came down from Enfield but nobody else, and we had some fine people there.

Perhaps Pheazey and others learned that their insensitivity did not pay off. Certainly when it came to the bigger moves to Harlow and Paignton they were obliged to persuade families to make the change, informing them, arranging visits, dealing with problems. Yet hierarchical attitudes remained in so many ways. When going on holiday Chris Foulkes, manager of the valve division, told Pheazey that in his absence one of his people would manage the division. The idea of delegating downwards was new; before, one always delegated upwards, probably to Pheazey himself. At

Woolwich the welfare fund, built up from the sale of scrap wood and sanitary towels, was under the charge of the personnel manager, who could authorise loans up to £5. Above that, he had to explain the circumstances personally to the company treasurer to get authorisation.

In the structured organisation, it was not so much what you did as to whom you reported. Letters had to be signed on behalf of the department head, which caused a top-level problem for Leo Firnberg in writing to licensee companies:

> I had to sign over Spencer's name, much to the disgust of a lot of people. One letter I wrote with a copy to F. C. Wright signed by me for Tommy Spencer in which I said 'We do this'. When Spencer came back he called me up and said 'Why did you say we?'. At the same time F. C. Wright wrote and said 'Why didn't you say I?'.

Bill McMillan, the press officer, invited some journalists on a trip by letter signed for his boss, the publicity manager, Jack Read. When Bill met them at the station some of them greeted him as Mr Read. It was part of an antique way of going on. Desks for the company were still being made in the Southgate woodshop, which as in the aftermath of the Depression was glad to have solid work, even if it was not executed with a cabinet-maker's precision. On the same site the term 'assigned stenographers' was still being used for secretaries. A top one might even have that expensive luxury, an electric typewriter, which could also perhaps be found in a technical handbook section producing information for customers.

Fuddy-duddy the company was but it was also very human and personal in its relationships. Things got done not so much by system as because somebody knew somebody else, a friendliness reflected in the twee pages of *Standard News* with its profiles of Standard Types and Characters, parochial news of the Social and Athletic Clubs, pictures of weddings, dinners, and presentations for forty years' service. Long service, encouraged by the non-contributory scheme was a virtue and in 1954, at Behn's instigation, the Quarter Century Club was founded, with chapters for each location named after prominent people within the company. At the annual dinners members reminisced. There were individual families that could tot up more than a century of service and were grateful for continuous employment. If they had managed to put something by they knew it was all right in the Woolwich Equitable Building Society because Sir Thomas Spencer was on the board. He became the Society's chairman when he stepped down as STC's managing director.

Concern for people ran through the organisation. When the East Coast floods of 1953 cost the company money a cable 'SERIOUS FLOODING NO LOSS OF LIFE' was sent to New York, which replied 'DELIGHTED TO HEAR NO LOSS OF LIFE'. Unlike some of its competitors it was not a hire-and-fire company. It commanded a loyalty from its people although it was never renowned as a great payer. Jack Williams, the Woolwich personnel manager, was put in an unenviable position:

> Pheazey and Spencer were both mean. On two occasions I can remember being told to cut half a crown [12½p] off all my recommendations. The argument was that you weren't dealing with your own money but the company's.

Nevertheless labour relations were good, perhaps because the lack of change suited the conservatism of the work force. There were qualms among older employees when new machinery was introduced, as in the coaxial cable shop at Woolwich, but even if people never got entirely used to it there was no cause for dispute. Strengthened trade unions, which had gained a second week's paid holiday from the Engineering Employers' Federation in 1951, were on the whole happy to accept the *status quo*. It was upset on two occasions at Woolwich, when employees struck for a few days. The first was in support of an AEU convenor, who had taken time off for union business, and the second over an increase in the price of a cup of tea off the trolley from one penny (less than ½p) to three half pence (just over ½p). Support for the sacked convenor soon fizzled out but the unions stuck out on the price of tea and the management had to climb down.

A sensitive issue that did not immediately become a problem was the recruitment of coloured labour. Jack Williams could only get immigrants and, realising that mixing people of different cultures in an old-established factory like Woolwich could cause all sorts of problems, asked Frank Johnson in head office what company policy was. Johnson's attitude was not to anticipate the reaction but to try and see what happened. It worked better than expected and the number of people from India, Pakistan and the West Indies soon became about one-third of the labour force. Race relations problems were not so much between white and

13 During World War II, powerful radio valves still had to be finished by hand. Here, an employee strives for perfection of shape and a balanced texture.

14 Alec Reeves of Standard Telecommunication Laboratories, inventor of pulse code modulation, demonstrates its basic principles.

STC

15 This proudly-posed crew of sixteen men (top, facing page) was needed to lay telephone cable between Birmingham and St Albans in 1914.

16 & 17 At Oxford, in 1957, (facing) a smaller crew did the same job; while paying out optical fibre cable between Hitchin and Stevenage in 1976 (above), manoeuvring the whole reel needed only two men.

18 At Southampton docks, submarine cable made in STC's adjacent plant is loaded into the hold of a cable-laying vessel.

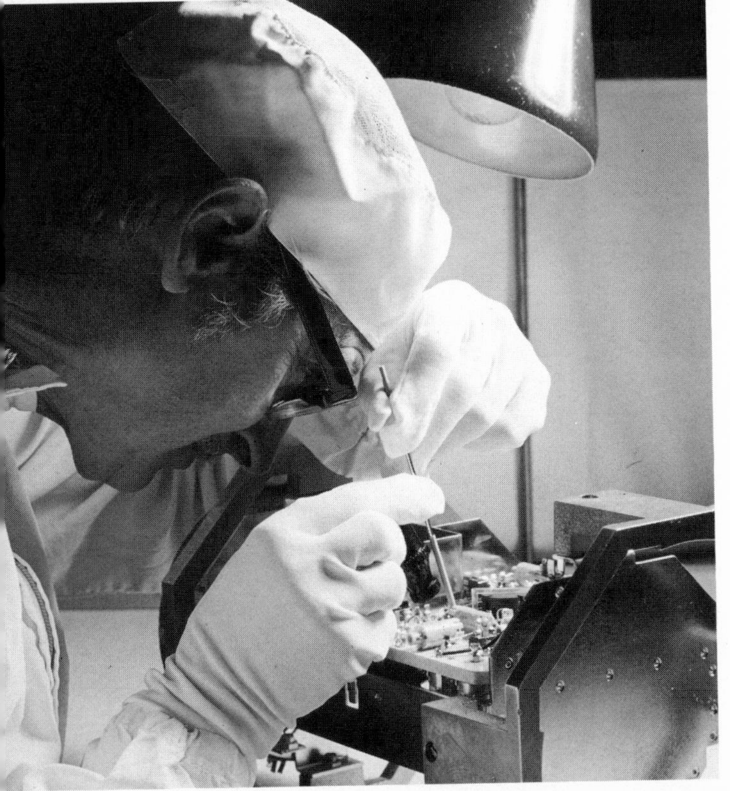

19 Surgically clean conditions are necessary in 'the dairy' at Greenwich, where repeaters are made to operate undisturbed for twenty-five years on the seabed.

20 The world's first undersea optical cable, manufactured by STC, being floated ashore in 1980 for its trial at Loch Fyne, a location chosen for its relatively deep, tidal salt water.

21 A TXE4A exchange is installed using STC's unique MITER system. MITER (Modular Installation of Telecommunications Equipment Racks) enables equipment to be assembled and tested at the factory. Once inside the exchange building, air cushions are used to move the units to their final position.

22 System X equipment in London's Baynard House exchange. System X – the fully digital switching system on which the modernisation of the UK network is being based – was developed jointly by STC, GEC, Plessey and British Telecom. STC withdrew from the project in October 1982.

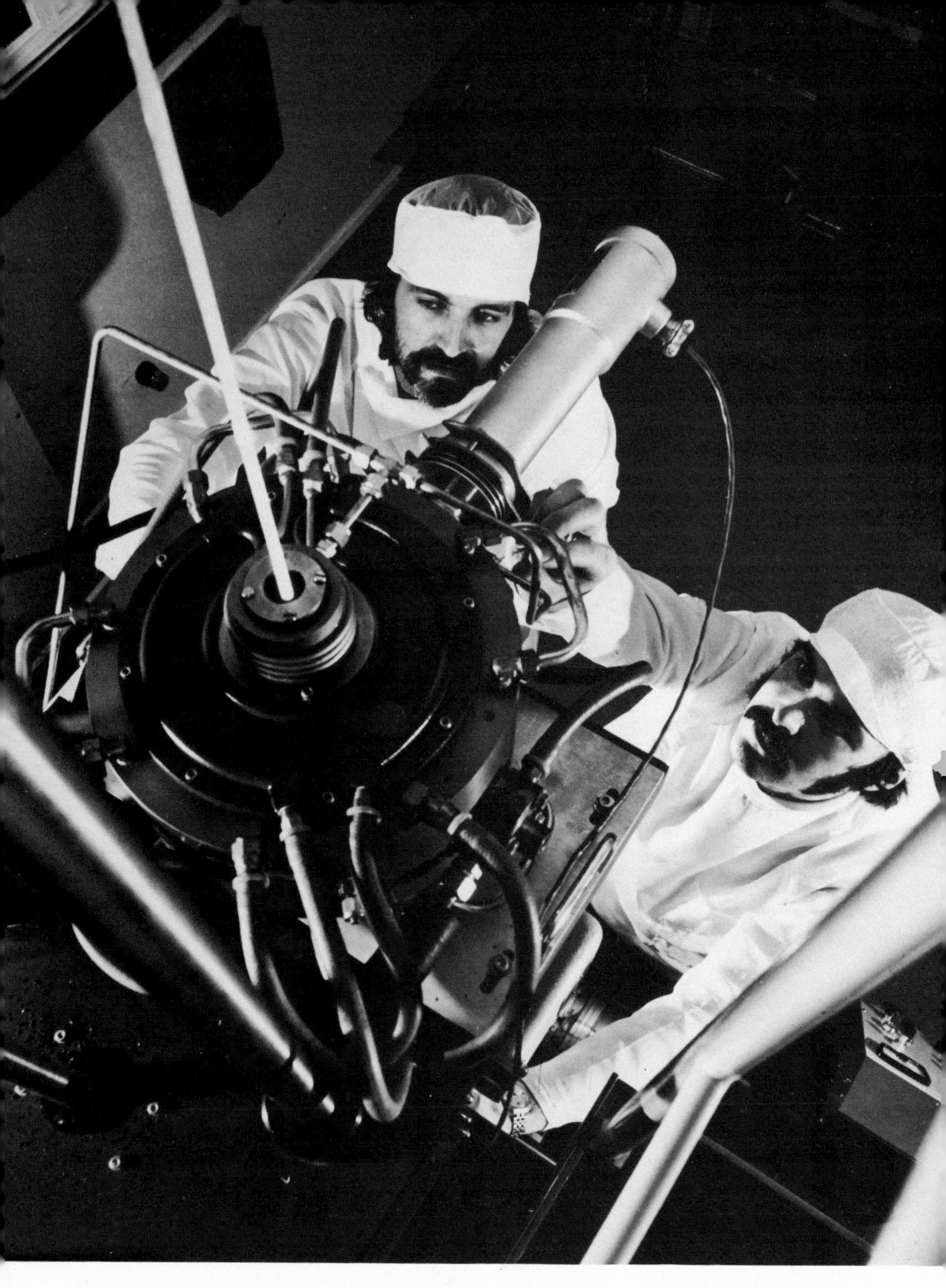

23 Optical fibres as fine as human hairs are produced at STC's optical cable manufacturing unit at Harlow. One end of a glass preform is introduced into the pulling tower, heated to melting point and drawn under steady tension onto drums at the lower end.

24 J. E. Kingsbury, first agent and managing director of the Western Electric Company in London. Though he relinquished an active role in 1909, shortly after this photograph was taken, he remained a director until the change of ownership in 1925. Up to 1915 he used his spare time to write the book for which he is best known, the authoritative *The Telephone and Telephone Exchanges*.

25 Sir Frank Gill was a telephone pioneer who started "down manholes and up poles". He became not only chairman of STC, but also an internationally respected figure in telecommunications.

26 Sir Thomas Spencer began life as a Woolwich lad with a passion for engineering – and football. He lied about his age to join Western Electric at 14 in 1907. By 1936 he was sole managing director, and embodied the spirit of 'The Standard' through the war and into the 1950s.

27 Sir Kenneth Corfield, who joined STC as managing director in 1970 and has been chairman and chief executive since 1979.

coloured as between different immigrant groups. White supervisors found themselves having to be circumspect in taking advice from one group to the disadvantage of another and more than once returned a £5 note put in with a Christmas card.

A problem the company did not fully realise was in public relations. Technically in the communications business, it was not in the business of communicating with the outside world. Jack Read joined as advertising manager in 1948 but, finding that the company did very little advertising, embarked on creating a radio feature with the BBC Overseas Service across the road in Bush House, in the absence of the commercial director prudently getting Spencer's approval before proceeding. Linguists were found in the plants, recordings made and a 20-minute programme assembled. When the final script was shown to McVie, the commercial director, on his return from South Africa, he exploded:

> Good God, you can't do this. We're not going to tell the competition what we're doing. What do we want to tell the public anything for? We're not interested in the general public. We're selling to the Post Office. We don't want this sort of publicity. It could get us into a lot of trouble.

Read explained the value of the programme, to no avail:

> What about ITT, McVie said. What about all our companies in Europe? What about the companies in France and Germany and Italy? When they hear this programme they'll think we're poaching on their territory.
>
> I finished up by having an awful argument with him and I went out of his office and he stormed out and I thought that was the end of my career with STC. I'd only been with the company for two or three months. Anyway, the following Monday McVie called me – I think he'd thought about it over the weekend – and said that he was very sorry that he went off the handle like that and would I go and see him. In the conversation I happened to say: 'I did clear this with Sir Thomas Spencer'. This of course calmed him down. Eventually the programme went out and it was really the very first time that STC had publicized its activities outside the country in that way.

Jack Read, who moved on to become publicity manager, found that he had won a battle but not a campaign:

> One had to struggle very hard indeed to get a press release out. One would write a press release and it had to be vetted with a fine tooth comb to make sure we weren't saying anything to the competition which was going to benefit them and we were not saying anything which we didn't really need to communicate with the public. After a while the company realized this sort of publicity wasn't doing any harm. It was doing us a bit of good.

Press releases tended to cover large contracts without mentioning the price but hinting that it was big, participation in exhibitions, and obituaries. No financial information was forthcoming.

The kind of public showing the company did make affected its recruitment and in the longer term the calibre of its staff. In the mid-1950s it was recognised that:

> The company must at once take steps to improve the grade of scientific staff it recruits. Today the company is able to fill its more responsible positions from people who entered its service a generation ago. Recruitment at that time may be divided into two broad classes, viz the college graduate and the schoolboy . . . It cannot be expected a generation hence that the non-graduates recruited today will yield anything like the percentage of capable men that entered the company similarly 30 years ago: boys with brains have no difficulty in getting State scholarships and their approach, if any, to the company is as graduates. Our experience at Ilminster is very definitely that the grade of college graduate that has been recruited since the war is considerably below that which was recruited between the wars; in fact, well over half the graduates engaged now would never have been even considered before the war. The outlook is therefore very unpromising. This is not surprising. Government service offers more security, more definite short-term prospects, higher social status, a much better pension scheme, longer holidays, perhaps shorter hours and generally more glamour.

One engineer put it more cogently:

> We are running on the 1914 to 1924 vintage, which is our backbone. We have no second line, and in ten years' time we will be out. We cannot get replies to our advertisements. It is not a question of money. Those people that do come are off within one or two years, because the outlook does not satisfy them.

Bill Dansie was personnel manager at Southgate:

> The switching technology was too stable, too long-lasting to be a great attraction to young men. Other forms of communication, e.g. radio and TV, were moving much faster than telephony. Computers were also coming along in the Fifties. Our professional engineering supply was on a closed loop. You either recruited from the telephone industry or you did not. We had poaching agreements below managerial level. So we wouldn't recruit a circuit engineer from GEC or Plessey without 'consultation'. This lasted until well into the Sixties. The effect was stultifying. Little new blood, new ideas. In our specialised industry circuit engineers of interest to us could only be found within the industry. Stagnation.

Southgate was near the site of a cemetery and it seemed that some of the dead had risen and were continuing to make the type of telephone exchange devised by their undertaker patron, Strowger. Even in the more lively radio division there was something of a government department atmosphere in that pay scales were compared to opposite numbers in the scientific civil service.

Relationship with the labs on a personal level was remote. The labs recruited their own staff, who worked on a research project until it was ready to be handed down to a product division for development. At this stage lab staff were likely to spend time on transferring the technology but there were not many prospects of leaving mundane company work for the retreats of research. Within the company it was not easy to move. There were instances of division managers refusing to let people go and the only way they were able to transfer was to leave and spend some time with another company before coming back. Throughout the industry the company was regarded as having a good apprenticeship scheme and being a good place to train. After that you either left with a good background for better things or waited for dead men's shoes. People hung on to a job because they felt it was a deserved inheritance not lightly to be given up even for the love of the business. Joan Smith saw the results:

> There were people who came into the office every day and it wasn't apparent that they were doing a great deal of work. Yet they seemed to be able to hang on to their positions. There was need for thinning-out of staff, reappointments and even cancellation of some jobs that existed. We did have over-

> manning. People who were perhaps good at a particular sport and had won the company acclaim through their prowess in the early days were able to hold their jobs, which seems totally wrong. The training was good but the prospects weren't visible enough. In those days you could definitely be in somebody's favour and get on. Or you could work very hard and not get anywhere.

If you did stick it out and eventually got to the top you had your garden tended by the works' staff. If you were a more ambitious young engineer you did not have this kind of horizon. Prewar arguments about the company's career ladder, the value of security in a steadily expanding industry, the non-contributory pension scheme being a form of supplementary pay did not carry much weight against the immediate prospects of more cash in hand and wider experience.

In the closed world of STC there was no management training. One man went to the Harvard Business School but he was an exception and he came back confused. People were promoted who were not fully competent or trained for the job and they had to learn as they went along. There were not enough people of the right calibre to deal with the problems the company had. In the middle of the decade it had been slow to respond to new competitor challenges in export markets yet it was diversifying into things like computers without having the talents to understand the different businesses, evaluate projects and control expenditures. At the same time the company was not investing enough in its traditional products, was making do with obsolescent machines. Crowded workshops were in need of modernisation. Above all, a completely new attitude, nothing less than a cultural change, was needed to running a business in the latter half of the twentieth century. With hindsight, people were to argue that the company was going flabbily downhill like AEI, which was taken over by GEC, or heading for bankruptcy like Rolls Royce. Sheltered from the stock market, STC could be saved from such a fate by only one organisation, its owner, ITT.

In New York what was to be a momentous change had taken place at the top. During the caretaker presidency following Behn, the board had headhunted a new president, Harold S. Geneen, a man who could have given his name to Hal, the computer in *2001*. Born at Bournemouth in the UK, Geneen had been taken to the USA at the age of 1, had been a good son to his deserted mother, started as a page on the New York Stock Exchange, studied accountancy at night and worked his way up by moving from one company to the next. In his restless mobility he developed a mastery of figures, reading balance sheets as others did detective stories, remembering numbers, seeing their relationships and plots unfolding, not guessing the outcome but predicting it. To him, business was not about making products but about making money.

It was not until he was hired from Raytheon by ITT that he was the man at the top, with the full authority to put his ideas into practice. When he took over he promised to double turnover and profits within five years. To do it, he quickly changed the character of his world headquarters, which was 'little more than a mail drop'. Henceforward 'management must manage', which meant setting targets, measuring performance, taking action so that results would be 'as planned'. It was 'a whole new ball game' and the gentlemen in STC were not very good players.

13

The Old Order Changes: 1959–1970

When F. C. Wright took over from Spencer he made changes in organisation and personnel, strengthening the divisional structure and making managers accountable. Once a month they came up to Connaught House to face the executive. Hitherto the only figure that had mattered was sales. Wright's innovation was an open and systematic review of performance, not as thorough as it was later to be but a beginning. He reinforced the central commercial organisation with new appointments, which also helped reduce the influence of the old guard. Personalities and attitudes were not going to change overnight though, as Geneen soon realised.

The news of his appointment in 1959 had reached Connaught House by telegram but it did not mean much. Geneen was just another successor to Behn and the available directories were not very informative. When he appeared he was physically unimpressive, shorter than the Colonel and a bit round-shouldered. Initially he seemed affable, almost cherubic, meeting people, visiting plants, asking questions. More inquisitions were to follow in his wake. The first shock was that the top people including the managing director, who had shown himself a better speaker than Geneen, became guinea-pigs for headshrinkers under the direction of an American vice president, Dr Hilton, a man in his 30s and part of the new regime. They all had to submit to three-hour personality and IQ tests, a group test and finally an interview. Service and status counted for nothing. Frank Johnson, the chief personnel manager, who was told to hire professional psychologists to carry out the programme, was not in favour:

> All this was resented like hell. We had not been subjected to anything like this ever before. I made the mistake of offering

some critical observations, which were not in my career interest . . . I'd been brought up on cost reduction, even to the extent of telephoning. One day I had a call from Hilton in the States: 'Oh Johnnie I'd like you to do this, this and this.' He went on for about a quarter of an hour on what he wanted me to do. 'How's the programme going Johnnie? . . . Keep pressing, keep pressing. I'll put all this in writing to you.' . . . The programme went through with some resistance.

One of the findings of the psychologists was that too many of the senior people thought of retirement rather than the job in hand. Only two men did well and the rest were demoralised. They also had to spend time explaining their *modus operandi* (which they irritatingly cloaked in mystique) to American consultants Arthur Andersen, put in by ITT to look at the structure of the company. STC itself, against some opposition within its own ranks, had also called in consultants, the UK company Production Engineering, on work study and the needs in individual plants. Many loyal servants of the company felt that they were forced to spend their time trying to communicate a lifetime's experience to outsiders instead of getting on with the job. A lot of the consultants' work was felt to be naive, without an understanding of the products and their particular market, an elaboration of what the new masters wanted to hear. What the consultants appeared to be doing was spinning their investigation out and finding convenient omissions and inadequacies of performance to create openings for themselves and their friends. If it was at the expense of the old servants, a number of whom like Pheazey and Helbing conveniently retired about this time, then in the new view it was simply evidence that they did not measure up to the professionalism of a tougher world.

F. C. Wright's old division, transmission, was 'a problem area' and criticism of it was an implied criticism of him. An American consultant evaluated his findings:

As a result of organization weakness, development and engineering delays have been numerous, fines and penalties have been imposed due to lack of complete engineering and testing, engineering morale has been affected and division performance has sharply declined. Although much needs to be done throughout the division, strengthening of engineering is one of the most important single factors if return to profitable operations is to be achieved.

In particular, a large project in Spain had gone wrong and the equipment had to come back to be modified. In another division there had been a much publicised, ill-fated venture into Auto-Stackers, a company that was to build automatic car parks. The only installation, on the site of Woolwich Empire, was a farcical turn worthy of the music hall – the cars had to be manhandled – and an eight-storey reminder that it was better to stick to the business one knew.

Dissatisfied with the performance at large and in detail, Geneen wanted his type of professional manager to get hold of the company, put it right and improve its profitability. It compared badly with the German company, as he pointed out to the chagrin of the British in the monthly review meetings of ITT Europe, the new co-ordinating organisation growing in Brussels. In the absence of F. C. Wright on an extended overseas visit and his recently recruited number two Dr J. D. Stephenson, who was on holiday, Geneen decided that an American would have to be put in. The director of manufacturing in ITT Europe offered to put STC on its feet 'in six months'. 'Done' was Geneen's immediate reply and Rex B. Grey, a short, tough, 40-year-old Texan who had once been a test pilot, moved into Connaught House. He was very different from the last American managing director, H. M. Pease. Grey put the pressure on: 'Get them up from the South Coast for an eight o'clock meeting.' He would pilot the company Dakota to an airfield near a provincial plant and the staff would not knock off at 5.00 p.m. If he felt like it he would hold a late-night meeting, Geneen-style. Face-to-face discussion was the way he preferred to conduct his business, often remaining silent so that the other man talked himself into a hole. Having little time for the niceties and delays of exchanging memos, he scribbled curt replies on the bottom stating the action to be taken, usually pronto.

The whole tempo of the organisation changed. Much of the business seemed to be carried on by international telex. Under Geneen, ITT was a new kind of corporation made possible by two inventions: the electronic computer and the jet engine. Computers kept track of the mass of figures, the data by which units were controlled and on which decisions were taken. Forewarned by trends in figures, staff could fly to potential trouble spots. 'I want no surprises', declared Geneen. Like the Russians before him, he worked to five-year business plans. Managers set down their expected performance for the next year in detail, in outline for the following year, and took more than a glance at the five-year horizon. It made them think about their business, assemble information and see it in some perspective.

When Geneen came to review it with them he made them think even more, and understand the financial implications and inconsistencies in front of their colleagues. The managing director was also exposed to cross-examination in front of his own staff. In the process they learned a new language: the meaning of terms like cash flow, return on assets, and key ratios. They acquired new attitudes: that they were businessmen rather than engineers, that there was no compulsion to be in a technology if it was not good business. In 1960 the Brimar domestic valve business was sold to Thorn and the company, unable to support the research and development costs, got out of computers. Fortunately the idea of closing the valve division, considered by some to be obsolescent with the advance of semiconductors, was dropped. Following government-imposed cutbacks by the electricity authorities, in 1960 too the loss-making power cable activities were merged with Enfield Cables and in 1964 Delta Metal acquired STC's holding in what had become Enfield–Standard Power Cables.

Resources were thus released for concentrating on the company's main business. Here Geneen had a vista of non-cyclical growth. Where Gill had seen the backwardness of Europe compared to America as a problem Geneen saw an opportunity. In the number of telephones installed per hundred of the population other countries would at different rates seek to catch up with the USA, the world leader in dynamic economies. People only had to get involved, get the message and be shown the way. It could be fun too. In STC Rex Grey was the first managing director to hold a Christmas cocktail party in his office for everybody, irrespective of rank. Starting at the end of 1961, he made an annual practice of getting the top few hundred people together and giving them the score. Bob Coles was one who appreciated the new management style:

> It was Rex's management information meetings that gave me a visibility of what was going on. He talked about profit, which had previously been unthinkable. Managers had to stand up and give the story of how their division was doing. They accepted the discipline of accounting for their unit, saying how it stood and showing it on charts.

At one meeting Grey had the company doctor handing out

pills at the door to those recovering from the night before, fined people who came late £1 not to be claimed on expenses, and appeared on stage in a bowler hat and carrying an umbrella: 'You know whose side I'm on'. In truth he favoured some of his countrymen and had little time for the effete Englishman who had lost his Empire and not found his role. Geneen addressed the meeting and the profits were pledged to him in a stage routine involving a sack of dollars and a huge imitation safe made by the head office carpenter. Some of the old guard resented the American razzle-dazzle but most found it gave new purpose to the company. At head office Bob Coles was responsible for marketing methods:

> It involved looking at whether training was being done properly, whether the organization of departments was as it should be, whether the ratio of marketing expense to sales was right – theoretical aspects of marketing. It was aimed at improving the performance of the marketing people in the divisions. RBG used his staff. He would ring up people on the HQ staff and give them an instruction to get something done in a division. As a result middle managers in HQ found themselves giving orders to divisions, saying 'Rex Grey wants this done'. You had to fight through the objections and see that it was done. It gave a far greater responsibility and feeling of achievement to staff people.

In Connaught House Grey had a team of functional directors, new men about his own age responsible for the main functions: financial, legal, marketing, personnel, production, and technical. He encouraged competition among them in getting up in the mornings and coming in with new ideas. Pat Elliott, the personnel director, introduced psychometric testing, which he did not import from the USA but out of his own conviction. Previously employee selection had been a matter of personal judgement. Elliott brought in tests that revealed personality traits and enabled interviewers to direct their questions to match people more closely to job specifications. They justified their procedures with arguments on reducing the ratio of mistakes and the economic benefits of not appointing the wrong person. The higher up the scale you went the more people a bad decision could affect and the longer it could take to put right.

In the new decade more outsiders were coming in. The reorganisations threw up a demand for people and talent that could not be found within, at all levels. There were jobs that had never been heard of before, functions like industrial engineering that had been carried out in a fragmented way under the title manufacturing planning. Through management consultancies, new managers were recruited and, although not regarded as real engineers by the STC management and as unwelcome innovators by the unions, they got down to making nuts and bolts changes on the shop floor, improving production processes. Changes were not always easily implemented, especially at Southgate, described by one of the newcomers as 'the graveyard of innovation'. It was difficult to get the technique of work study accepted let alone alter the manufacture of a product that had been going on for forty years under a paternal management. One industrial engineer had proposed a change in the manufacture of a Post Office relay the week that his daughter was born. She had gone to school before the change was implemented.

Piecemeal and indirect approaches helped. Quality assurance departments were set up. The company's first industrial designer, Richard Stevens, was appointed. His approach was not to style a product as a cosmetic process but to make its appearance an integral part of its function. For example, he redesigned test instruments produced at Newport, which had a traditional robustness as though they were meant to operate down neighbouring coal mines for fifty years without dent. In 1963 the company launched the two-colour 'warbling' Deltaphone with a lightweight body only slightly wider than the dial itself and a handset about half the weight of previous types. The instrument won a Council of Industrial Design award in 1966 and was marketed by the Post Office under the name Trimphone. Stevens later became the Post Office's own industrial designer.

Change was helped by technological advances, like the electronic tone caller instead of a bell in the Deltaphone. In cables, by 1960 plastics technology had almost entirely taken over and only a few rubber types were made. Transistorised transmission systems, smaller in themselves and used on smaller core cable, were coming in. New products led to other changes but they also helped to preserve the old attitude that development was something that happened instead of being consciously planned and scheduled. Yet under the Geneen regime that notion was being applied to the market, which was outside the company's control. Market research was done on Post Office requirements to get breakdowns of what would be needed in terms of cables, exchanges, transmission and other equipment. Not until the latter part of the 1960s did a similar discipline come to be applied to engineering development, with a case being documented for a particular objective to be achieved, for a purpose, by a certain date, within a budget. Then it was possible to evaluate one project against another. For years people had lived under a regime without an investment policy and they did not immediately think in terms of spending to improve.

Joe Hitchcock, who had a background in production control and was from 1964 head of the largest group, telephone switching, fought a rearguard action against the introduction of a computer system, which he believed would cause an upheaval and would be superseded. With attitudes like this at the top it was not to be expected that employees would always readily accept change. Much depended upon the place and the product and fortunately for the company its industrial relations were decentralised, based on plant bargaining. In 1963, when the Engineering Employers' Federation had concluded a three-year national agreement with the Confederation of Shipbuilding and Engineering Unions, STC wanted to take advantage of the agreement by introducing wage plans in return for co-operation on work study and changed methods. ITT held an investigation into labour relations practices in Europe. At a one-week conference held in a castle near Shannon the British had to explain that agreements with UK unions were not legally binding and that the only way to make progress was to establish mutual understanding and respect. In this way term agreements were secured in some plants, mainly within the components group, which had to make changes to succeed in a much more competitive market.

In traditional products tied to the Post Office things were not so easy, as Joe Hitchcock saw:

> One of the groups that perhaps gave more trouble than most was the cable formers, who could see that they were of a particular craft, had a very specialized skill in making cable forms for switchboards and racks, and that this could be coming to an end with the change in technology. They were always very jealous of their craft and we had very many battles with them over the years on status. We tried to automate, or semi-automate, cable forming and they resisted us very strongly.

The slogan adopted for the 1963 management information meeting was STC – Strength Through Communications – an attitude that was beginning to extend beyond that immediate gathering. One aspect of becoming a more open society was communicating more with the outside world, telling the press and local communities more about what it was doing and why. The company began to get closer to its employees. *Standard News* was renamed *STC News*, its format changed from magazine to newspaper and an editor with a shop-floor background appointed. Gone was the tweeness, but in many ways it was still his master's voice in another guise. Changing the editorial stance was another revolution that would not occur overnight.

Nevertheless, all the changes were adding up to a shifting outlook. Early on, Grey realised that one of the great assets of the company was the dedication of its staff, a loyalty that had to be retained and redirected. After the upheaval and demoralisation he brought some stability, speaking the language of ITT Europe – which regarded him as a maverick – yet protecting STC from too much interference. He might not be able to get the company on its feet in six months but to have it in step along a new path was another matter. A younger man, unconstrained by his background or personal loyalties within the company, Grey seemed to be accomplishing more than F. C. Wright could have done within the same period, but he could not achieve a sudden transformation.

External circumstances had changed in his favour. The effect of the Street Offences Act 1959 was to take prostitutes off the streets and put them on the phone, ushering in the permissive society of the Swinging Sixties. In the words of media guru Marshall McLuhan: 'The phone puts the call girl at everybody's disposal'. For other legitimate reasons demands were being placed on the creaking Post Office network. With rising living standards more people wanted to use the telephone and waiting-lists were growing, as was the number of failures to get through, especially on trunk calls. Questions were asked in Parliament and successive Postmasters-General blamed the worsening performance on consistent undercapitalisation. The Treasury had received all the revenue from the telephone service and allocated only approved expenditure. From 1961 the situation was improved slightly, with the Post Office getting more control over its finances but still having to secure Treasury and parliamentary approval of its investment programme. Although telephone development was planned over a five-year period and expected capital requirements known in advance there was no guarantee that public expenditure would not be cut.

On the technical side the Post Office believed that it could solve its problems with the obsolescent Strowger exchanges by going straight to electronics, a view shared by STC, which had rejected the intermediate crossbar system produced by its associates on the Continent. Under the Joint Electronic Research Agreement the manufacturers tried to produce an advanced switching system based upon the principle of time-division multiplexing. The result of several designs was the Highgate Wood exchange, handed over to the Postmaster-General on behalf of the industry by Sir Thomas Spencer, STC's chairman. A committee product representing the work of suspicious and competing manufacturers, it was short-lived, too ambitious for its period in terms of the available technology and components, and very expensive. Reliable transistors and integrated circuits would make electronic exchanges economically possible.

Meanwhile there was plenty of conventional work, little of it of public interest. In 1965 the company won the order for equipping the UK's best known telephone number WHItehall 1212, the new communications centre for Scotland Yard. Alas, the following year it became just another number when all-figure dialling was introduced. Microwave equipment providing 960 telephone circuits was supplied for the London-Bristol section of the new network radiating from the 619-foot Post Office Tower in London, and later extended to Exeter. The moulding division made souvenir plastic models of the new tower for sale by the Post Office. New work was also coming along, like the production of the Deltaphone for public networks and the push-button office version, Deltaline. In 1966 the Post Office placed its first orders for pulse code modulation (pcm) equipment, which enabled it to increase the capacity of existing underground cables between exchanges twelve-fold and improve the quality of transmission. Suitable transistors and, more recently, integrated circuits had made pcm practical in terms of size and cost. Ahead of his time, its inventor, Alec Reeves, was at last recognised: the Ballantine Medal of the Franklin Institute (1965), the City of Columbus gold medal (1966), and pcm on a shilling (5p) postage stamp (1969).

To cope with rising demand, mainly from the Post Office, the expansion of the latter part of the 1950s was continued. Investment in passive components like capacitors was maintained. In Harlow an electro-mechanical division was formed for the mass production of relays for the open market. It took over the factory of the quartz crystal division, which moved into a new plant where higher yield home-grown quartz was introduced. In three weeks it was possible to produce crystals that took nature 3 million years. The components people were moving away from being pulled by the systems divisions and, as a coherent group, were establishing their own identity, pushing into new markets. They suffered a setback in 1964 when semiconductors were put into a world-wide group but behind them were internal wrangles about dealing with third party customers and the doubts of those customers that their equipment specifications would be given to their arch competitor, STC.

Having lost business with the cessation of some of the company's own activities like computers, the group had to find new markets. Along with the increased output of components went a new distribution centre, STC Electronic Services, which offered a fast supply of its own and other companies' products. Started in a small warehouse at the back of the rectifier division, originally to market an equipment practice with components as a secondary line, it was to grow quickly, encouraging competition. Its success, not foreseen by Grey, who wanted to close it down, depended upon the fact that fewer companies could afford their own direct selling organisations.

At Woolwich, transmission work was being carried out in the cramped conditions of an old plant. To relieve them, the development and engineering labs of the microwave division were temporarily moved to St Mary Cray, Kent, a place more likely to attract bright young engineers. In 1965 the labs moved to join the greater part of the transmission systems group in the new £2¼ million 450,000-square-foot factory at Basildon, the other new town in Essex. The largest factory built for the company since the war, it was opened by the Postmaster-General, Anthony Wedgwood Benn, who noted the old alongside the new: small batch production to customer requirements in a random pattern; and manufacture of standardised units as building blocks by flow-line methods. A sign of the affluent times was the seven-acre employee car park with room for 1,300 cars.

The stimulation of demand proposed between the wars had come. Installation of telephones called for more subscriber and switchboard cables from Newport, where a 50,000-

square-foot extension was built and the plant redesigned to produce more of the plastic types that had superseded rubber. There was further expansion in the Celtic fringe, mainly for the manufacture of exchanges and telephones. In a fairly settled Northern Ireland the first factory was rented in 1962 at Monkstown, followed by smaller plants at Larne and Enniskillen, which altogether brought employment for some 4,000 people. The commitment was big enough for a separate company to be created, Standard Telephones and Cables (Northern Ireland) Limited. Like other newcomers, it was an equal opportunity employer and sectarian feelings were not an overall problem. A tougher proposition was the Scottish new town of East Kilbride, where Clydeside industrial militancy persisted. In contrast, the converted water mill at Bovey Tracey in Devon, acquired to supplement capacitor production at Paignton, was a haven of rural peace.

To reflect the image of a growing, modern organisation the company moved its headquarters in 1965 from Connaught House in Aldwych, which became part of the London School of Economics, to a new ten-storey office block opposite St Clement Danes, the Wren church famous for its *Oranges and Lemons* chimes. Partly on the site formerly occupied by *The Illustrated London News*, STC had as its address 190 Strand, which had been the antiquarian bookshop of Peter and Helen Kroger, jailed for spying for the Soviet Union. To the company, which wanted to remain in 'the electrical area' near similar organisations, it was a convenient address. Unfortunately, after the decision to move had been taken most of the neighbouring electrical companies moved elsewhere. One of the innovations in the new building was a staff restaurant, which helped people to mix more easily.

At a time when there was concern about the 'brain drain' to the USA the company had succeeded in bringing back more scientists than it had lost. The numbers were not great but a 'brain gain' was news and an opportunity to point out that there were good research and development facilities on this side of the Atlantic, a point made in Parliament when STC was mentioned in the same context as ICI. At Standard Telecommunication Laboratories the emphasis of the work was tending to shift towards materials and microelectronics developments with probable applications in view and there were also government projects, some secret, to be seen through. Via technical committees there was liaison with the

European companies, especially the Paris labs, and exchange visits with Bell Labs continued. In 1967 a new wing was opened at STL.

Past research and development was paying off. Nowhere was this more apparent than in the radio division, which besides providing boosters to relay BBC and ITV programmes to areas of poor reception and continuing business in communications transmitters for radio stations at Leafield, Ongar and Admiralty establishments at home and abroad, had a growing output of aircraft communication navigation and landing equipment. Among this was a highly sensitive radio altimeter that could measure changes in height of as little as a foot, a degree of accuracy necessary in the critical moments before touchdown. Early aircraft to be fitted with it were the Trident and the VC-10. Another product, the instrument landing system localiser, gave accurate lateral guidance information suitable for automatic landing. In Autoland, the directional information fed into the aircraft by radio is converted into signals that operate the aircraft controls and the human pilot flies 'hands off'. The success of the technique was demonstrated in December 1962 when a number of 'blind' landings were made in thick fog at London Airport. Progress to fully automatic landing in passenger service from 1967 – there was an intermediate stage, Autoflare, in which at the final stage before touchdown the pilot still handled left–right steering – was gradual but at no time was there any doubt about the reliability of the radio equipment, which anyway was duplicated and even triplicated. Other radio division products of minor note were for the radio interference suppression and the notch no-drag aerial on Concorde and a similar aerial for the Harrier 'jump-jet'.

Success in this advanced technology led to orders for overseas airports, many of which were being modernised to handle increased tourist traffic. By making it possible for aircraft to land in conditions of poor visibility, delays, cancellations and diversions were reduced. Countries to which landing aids were supplied included Australia, Bulgaria, Egypt, India, Iran, Iraq, Ireland, Rumania, USSR, Saudi Arabia and Yugoslavia. When airborne radio equipment was supplied to Vickers for installation in Viscounts sold to China an international contretemps arose that cast a sidelight on the status of STC. There were objections in Congress that an American-owned company was supplying advanced equipment to a Communist country hostile to the USA, but STC stuck by the principle that it was British-based and supplying to a manufacturer in its own country.

In another technology, undersea cables, the company also scored international successes. TAT-1, the first transatlantic telephone cable, had changed the pattern of intercontinental communications. Until good voice communications were established in 1956 revenues from record services well exceeded those from the telephone service. With the overnight improvement in the quality of reception, telephone traffic soon became the major revenue earner and TV material could be transmitted on a slow scan system. Not wanting to be left out, the French established a 48-circuit link to Canada in 1959, for which STC supplied nearly half the total cable and the repeaters on the Newfoundland–Nova Scotia section. In 1961 a third link across the Atlantic, CANTAT-1, from the UK to Newfoundland, was entirely supplied by STC: 2,072 nautical miles of lightweight cable and ninety repeaters providing 80 circuits. The lightweight cable, which did not twist under tension, with minor variations was the type that would be used on every major undersea cable system in the world. Whereas TAT-1 proved undersea technology, CANTAT-1 showed the way it was going to develop.

Western Electric in the USA had developed an alternative design of lightweight cable, which AT&T wanted to use with its own repeaters in the first link to Europe without an intermediate landfall. The cost would be shared with the Post Office and it was decided in 1960 that the repeaters would be American made and the cable British. Gordon Duddridge was involved in the negotiations for what was a crucial contract:

> The intention on the part of the Post Office was that the cable would be made between Submarine Cables and STC. A lot of time was spent at meetings trying to get a cost investigation basis and some system to enable the companies to regain or to cover part of the plant costs in case the business was never repeated. The only thing to do was to go all out for all of it. After a long series of meetings in London we got the whole TAT-3 contract. Submarine Cables got nothing. The decision was on a straight commercial basis. There was a single order for 3,518 miles of coaxial cable, which involved building a new factory specifically to the requirements set by the customer. We had to get the factory, our number two factory, running within twelve months of starting. We financed it in part by getting from the customer the difference in the cost of

> the building and what it would be worth on the open market if we had to sell it afterwards. We hired the machinery from a finance company. There was £2 million worth of machinery and about £800,000 of factory. We had that going in 1962 and that trebled our capacity. Our place was twice the capacity of Submarine Cables Ltd. From that time, the time of our success with the American business, Submarine Cables Ltd ceased to make any profit at all.

Winning the contract was also important in establishing a working relationship with AT&T on undersea systems. In disagreements with the Post Office, which always had a closer relationship with the British-owned Submarine Cables, STC found itself more often on the side of AT&T, from whom the real impetus for transatlantic cables was coming.

One of the reasons for setting out to be big in this type of business was that it tended to be feast or famine. As the number of orders was relatively few it paid to be able to take as big a share as possible. When business was slack, as it was for a year or more in the mid-1960s, the machinery was applied to extruding a slotted plastic pipe for land drainage, but it was a poor substitute for profitable export business in large contracts. There were systems in the North Sea, the Mediterranean, the Caribbean and the Pacific that earned the company a 1968 Queen's Award to Industry for export achievement. It could also boast the world's first 360-circuit deep water installation, between South Africa and Portugal, code-named 'Greenland' – checking with the company's press office when the ill-kept story was going to break the *Financial Times* technology correspondent used to sing 'From Greenland's icy mountains . . .' down the phone – and the world's first 640-circuit system, from the UK to Portugal. By the end of the decade an 1,840-circuit system was in production.

Advances in cables were stimulated by progress in space satellites. STC had only a marginal involvement in these developments and in 1966 teamed up with Vickers to bid for ground stations. No more came of the partnership than of the two companies' scheme for a series of manned transatlantic platforms, like North Sea oil rigs, connected by undersea cable to provide air traffic control information. The need was later met by satellites. STC had to make do with orders for radio equipment: a transmitter for the Post Office ground station at Goonhilly Down and some lincompex equipment for improving the speech quality over long-distance high-frequency radio links to twenty countries.

There was business in conventional products: radio equipment for Chile, Colombia, Panama and Peru; following the first international broadband link between Folkestone and Loos, microwave for Brazil – a contract the company later had to negotiate its way out of – Greece, New Zealand and Spain; multiplex equipment for Malaysia; the first 2,700-channel coaxial cable system in Sweden and greatly increased capacity links between Austria and Italy; a £6 million order for tank and vehicle radio sets for the West German defence ministry; and a control scheme for the Iranian Oil Refining Company. The more technologically advanced a product was the greater chance it had in export markets, even though the orders might be small, like the first export of pcm to Finland, the Deltaphone to Kenya, and the first solid state 2,700-channel coaxial system installed in Austria, which won a Queen's Award for technical innovation.

With the exception of Colombia, it was only colonial administrations previously dominated by Post Office practice that were interested in the Strowger exchange, and they were declining in number. One of the gestures of independence was to break away from economic and technical apron strings. Being tied to Post Office business presented another problem disturbing to a marketing man:

> When Post Office demand was high we didn't want to know about export. When it was low we tried to get in on export on a short-term basis. We turned the business on and off with the Post Office ordering cycle. It takes two to three years to persuade an export customer to buy a system, by which time the home market might well have changed. You haven't the production capacity to meet export needs. We went through this cycle several times.

To gain industrial orders the company was having to adapt from the attitudes that went with rigid Post Office specifications and meet the needs of individual customers. For example, it supplied an electronic reservations system to BOAC, using a computer from its German associate and its own terminals. Over CANTAT, seat availability could also be checked in Montreal. To handle the real-time data systems market a new division was formed. Its principal activity was marketing an automatic data exchange (ADX), a computer-based successor to the STRAD wired-programme system for routeing telegraph messages through networks. Initially most of the market was among military and government organisations that had high volumes of traffic. An ADX was installed, for example, at Shannon Aeradio Communications Centre to speed up message handling for North Atlantic air traffic. Later users of these large systems were newshandling organisations like the BBC and Reuters. Data terminals were supplied to Time-Life for sending information between London and New York and a three-nation hook-up demonstrated involving London, Paris and Washington. It was the sort of occasion on which a transmission engineer like Kenneth Hodgson could look back and see real progress:

> In 1934 I bought a new Standard 9 car for £140. That was roughly the same as a Mini in 1965 costing £360. Over the period the money price increased by about 2·5 times but the value of money had dropped by about five times, so the real cost was halved. Between the same dates the selling price of a channel modem was reduced by about 5 times in money or 25 times in real terms.

Another means of communication into which the company expanded – it had already bought the small company Hudson Electronics – was mobile radio, recruiting for the purpose John Brinkley, the managing director of the market leader, Pye Telecommunications, and the man who at the Home Office in the 1940s had been responsible for the development of police radio in Britain. Under him, a mobile radiotelephone division was set up to make and market a newly designed range of UHF sets for installation in vehicles and later as pocket sets. Winners of a Council of Industrial Design award, they were used by airlines, British Rail, on London buses, by construction and industrial companies. Also in radio the company marketed a remote control system for electric overhead travelling cranes so that operation could be carried out from the floor instead of the cab. Selectronic was an updated and faster means of supervising electrical power systems. To handle miscellaneous business a project and field services division was formed, which among other things took the responsibility for a £500,000 postal mechanisation contract.

The company had also been expanding through small acquisitions. In 1961 it acquired Phoenix Internal Telephone Systems, changed its name to Standard Telephone Rentals and made it part of the private communication equipment division, what had been the heterogeneous industrial supplies division. Under a new manager the division had gone back into a 'ring' for PABXs and was getting its allocation of bread-and-butter business. In consumer products, to boost the small market share of K-B, STC acquired Ace Radio of Rhyl, Regentone, RGD (Radio Gramophone Development Company Ltd) and Argosy. These lesser known brand names soon disappeared and for the time being the company was not one of the few really big consumer electronics manufacturers. It tried a few gimmicks like a home juke box and combining a TV set with a record turntable but without success. In the growth rental market, where fortunes were being made, its operations were confined to a few areas in the north and, based on slot machines to compete with suppliers of clothing against weekly payments, did not enjoy a high reputation. More successful was the purchase of a family company older than STC itself. Robert Maclaren of Glasgow had started in 1844 by making chemical footwarmers for travellers to joggle on the new long-distance trains and when steam-heating came in the company had started making thermostats. For STC it marked the beginning of an involvement in the business of heating, ventilating and air conditioning.

Through ITT acquisitions, STC was also finding itself associated with all sorts of companies that had nothing to do with telecommunications. The traditionalists were shocked at the news of the parent buying a car rental outfit like Avis or a tap or pump company. In a conglomerate of interests from insurance and hotels to dog food and perfumes, under a set of initials that no longer represented what they stood for, old-timers averred that there had to be a reduced commitment to the business that was the foundation of it all and that they knew thoroughly. STC, hitherto about one-sixth of the population of ITT, was being diminished in stature. Competitors were bound to exploit the situation. Seen from this angle, corporate policy of diversifying interests and achieving a balance of earnings between overseas and the USA, where it was very difficult to expand in telecommunications, had little relevance. For the time being though in the UK, where mergers within industries were being encouraged, the blurring of the company's identity was not a serious threat. The dynamism of an expanding economy was sensed by the company secretary, John Buckland, who had joined just in time to shovel coal on the wharf at Woolwich during the general strike of 1926:

> I felt a lot of the excitement of business coming back . . . One

> Friday I was telephoned at about ten o'clock – could I sign some cheques for one of the subsidiary companies that I was secretary of? I said 'I'm terribly sorry I'm just off to the House of Lords.' 'There's no other secretary,' he said, 'and I must have these cheques to pay the wages.' I said 'I tell you what, I'll meet you outside Westminster tube station. There's a newsagent's stand there and in two or three minutes I'll sign the cheques.' And so that's what I did. I gave the chap a couple of bob [10p] for using his stand.

Some of this excitement was being shared by employees who, no longer under a paternal management, were getting encouragement to come up with ideas and were better rewarded for their contributions. An improvement in the technique of cable-jointing earned one man a Ford Escort, and four-figure cheques for suggestion awards, based on a percentage of savings, became increasingly common. Employees of all grades, rewarded like Stakhanovites under the Russian five-year plans, were eligible for an extra week's free holiday for two at the company's Costa Brava ranch house. Flying in the company Dakota was not exactly luxury but at least they could feel they were taking part in the world of international travel in which executives regarded plane trips as lightly as bus journeys. It was an aspect of Harold Wilson's 'Britain that is going to be forged in the white heat of this revolution' in which there 'will be no place for restrictive practices or outdated methods on either side of industry'.

Edward Heath in ending one restrictive practice, resale price maintenance, had helped strengthen public opinion against bulk supply agreements. The Committee on Public Accounts repeatedly criticised the arrangements and those for cable and transmission were discontinued in 1963 and a modification made in those covering telephone apparatus and exchanges so that a fixed percentage of equipment could be bought outside 'the ring'. In theory 25 per cent of telephones and 10 per cent of exchanges could be purchased outside but in practice, especially on exchanges, it made little difference. John Brinkley of Pye, which for years had fought bitterly to get into the telephone ring, wrote to *The Times* on 24 June 1964:

> . . . the present arrangements by which the Post Office purchases its telephone equipment are not in the public interest. These Post Office Bulk Supply Agreements are amongst the most restrictive documents I have ever seen and it is

> impossible to reconcile them with a modern competitive industrial situation.

The Post Office Engineering Union published a brochure, *The Telephone Ring*, pointing out that British exports of telephone equipment had stagnated. A telephone dial made by the ring was reckoned to cost 26s (£1. 30), one by an outside firm 21s (£1. 05), and one from the Continent 15s (75p). Had purchases been made from the Continent, the supplier might well have been ITT. At home the POEU argued:

> Telephone rentals and charges for calls are quite high enough and are an obstacle to the wider use of the telephone . . . The public ought to know whether these are due to the present methods of buying telecommunications equipment by the Post Office.

The ring put up a defence of its relationship with a monopoly purchaser but it was going through the motions before the remaining agreements expired – telephone apparatus on 31 March 1968 and exchanges on 1 October 1969 – when the Post Office became a public corporation.

In the meantime STC's management had changed. Rex Grey, the American managing director, who had taken to South Africa during the negotiations on the 'Greenland' cable and had bought a ranch in Rhodesia, was interested in setting up a new regional organisation for ITT covering Africa and the Middle East. In mid-1965 he became chairman of STC, being succeeded as managing director by the financial man Alister Mackay, with Tommy Spencer, 77, becoming the first honorary president of the company. At this period, to present ITT as a group of national entities, Ellery Stone was recruiting to local boards distinguished men like Paul Henri Spaak in Belgium and Trygve Lie in Norway. In the UK he approached Harold Macmillan, whom he had known in the war and later lobbied as Minister of Defence over the 'Deep Freeze' cable. Macmillan turned him down and the seat was taken by Stone's wartime political adviser in Italy, the recent head of the diplomatic service and now provost of Eton, Lord Caccia. F. C. Wright, the former managing director, eclipsed under the Grey regime but valuable for his relationship with the Post Office, gained a CBE and was rehabilitated as deputy chairman.

His negotiating skills were needed, not only for the emerging competitive period but also for the new situation created by a sudden reversal of Post Office policy. Since the failure of the Highgate Wood electronic exchange in 1962 the Post Office had to cope with the problem of satisfying demand until a workable electronic solution could be found. To help quell public protests at the shortcomings of the service, it decided to adopt the interim solution it had always rejected: crossbar exchanges. Plessey, having developed and successfully exported this type of exchange on its own initiative, was quick to take advantage, which made STC panic. Pressed by ITT, which could not understand why the Post Office was the exception to the European pattern, and anxious not to be left out, the company hired some highly-paid Swedish and competitor engineers to re-engineer the proven Pentaconta system, established manufacture at its most troublesome plant, East Kilbride, and took export contracts in Peru and Zambia. Disaster. Pentaconta, established in Europe, had to be re-engineered completely for the UK network, there were labour problems (comparison with French production rates did not help) and management turnover at East Kilbride, and the prices were wrong. The two loss-making export contracts were the price of demonstrating to the Post Office that the company had a system. Nor was the home price right and it took long and heavy renegotiation to get it to a realistic level.

On top of this was a new threat in this type of business. At the beginning of the decade two companies, Automatic Telephone and Ericsson Telephones, had been taken over by the aggressive Plessey, thus reducing the big five manufacturers to four. In 1967 when Arnold Weinstock of GEC in his crusade to reorganise Britain's lumbering electrical industry made a bid for AEI, part of that company's defence was to link its profitable telecommunications interests to those of STC. Weinstock, a hero in the City, won and it was now the big three, except that STC looked vulnerable. On the present allocation it had only 20 per cent of the exchange business to 40 per cent each for GEC and Plessey. Moreover, although it could claim several contributions to the British economy, it was the odd one out in being foreign-owned.

To make matters worse, within the company the formerly successful radio division was losing money. On 18 November 1967 the pound was devalued from $2. 80 to $2. 40 and the 14·3 per cent difference made it harder for the company to achieve its net income targets in American terms. Whatever Harold Wilson might say, the pound in the company's purse was definitely affected. In sterling terms too, although sales were showing a healthy increase, net income was falling. None of this helped Alister Mackay, who took a nationalist line. He did not want to identify the company with ITT, going so far as to keep those initials off the front page of *STC News*, but ITT was reducing the empire under his control. For tighter, more professional management of specific products, international groups were being formed, in 1967 ITT Consumer Products Europe, intended to have a more coherent economy and a stronger brand image. It was followed more successfully in 1968 by ITT Components Group Europe, again promoting itself as a transnational body having all the strengths that implied. Mackay's policy, encouraged by the selective employment tax that militated against head office numbers in favour of productive people in factories, was one of decentralisation, pushing responsibility down the line and making the role of head office staff one of persuasion. ITT tried to persuade him to adopt its policies, imposing them where it felt necessary, as in the appointment of the company's first public relations director, Donald Grant from the Ministry of Technology, a ministry then important to the company. The crunch was financial. Mackay did not achieve the targets he set down in his business plan and was succeeded as managing director by John Ayres, who two days before Christmas 1968 announced his reorganisation.

Changing the reporting structure and making economies like closing the staff restaurant in headquarters and the ranch house in Spain were not enough to solve the company's problems, which were more deep-seated. A cost-cutting caretaker, Ayres pursued the decentralisation policy of Mackay when what was needed was a return of authority to headquarters and thorough reconstruction of the company. The divisions were pursuing their own policies, often underground. In a sense the Geneen revolution had not occurred in STC, which had been able to conceal its weaknesses during the boom of the 1960s. Economic difficulties towards the end of the decade had not caused so much as accentuated its problems. The internal resistance to Geneen, who had shown himself more than a passing president, had faded. Yet there had been more illusion than reality about change. Some thought that if the retired F. C. Wright, who had had a grasp of the problems and was developing his strategies, had been given the chance and had overcome the resistance and inertia beneath him, he could have put the company in a stronger position much sooner. Instead Geneen was promised and accepted a series of quick remedies when in the words of one man who was a candidate for the hot seat 'it needed a combination of Leonardo da Vinci, Jesus Christ and Einstein – and five to ten years'.

Had the company not been owned by ITT, it would have been, in the merger fever, a candidate for takeover. Through the Industrial Reorganisation Corporation the Labour government had been encouraging the creation of large industrial units. Tony Benn, the Minister of Technology, had discussed the situation at lunch with Geneen, who had opened with a ten-minute lecture on capitalism, to which Benn had replied: 'Oh, come off it. Surely you can't believe all that folklore.' A possibility was the nationalisation of STC, which would become a public foil in dealing with the two large private organisations *vis-à-vis* the new Post Office Corporation. However, Labour lost the 1970 election and nothing came of any such scheme. The company was saved but it was in no great shape.

14

A Quiet Electronic Switch: 1970–1979

Although the Post Office formally became a public corporation in 1969 it did not immediately change its attitudes. There was talk of a new era in which the former civil servants would become conscious of market forces, creating more competition among their suppliers and responding to the needs of their customers. The fact that the changes were delayed was one of the few advantages of STC's new managing director, Kenneth Corfield. It gave him time to reshape the company. When he took over in 1970 morale was low. After four managing directors within the last decade a typical employee comment was: 'Oh God, I suppose he'll change all the partitioning again'.

His approach was more fundamental. Having grown up in modest circumstances in the Midlands during the Depression he had seen poverty and felt cuts in his pocket money. He had also seen in the example of his grandfather, a pattern maker who established his own foundry producing quality malleable castings, that it was possible to create one's own future. At the age of 10, on his way to becoming an engineer, he had tried to build a Baird-type TV set in his attic laboratory. Later he established his own business making photographic accessories and cameras.

Having also gained management experience in large companies, he joined ITT in 1967 because 'it was such a big company that I didn't think I could run out of road'. His first major job was pulling together and building the European components group, the British arm of which was supporting the loss-making telecommunications activities of STC. Nevertheless, with all his achievements and experience, at 46 he was still something of the shy provincial come to London. Not far below his diffidence, however, was a determination to succeed, and when he said he was going to do something he meant it, an attitude not immediately believed throughout the company. A pragmatic Conservative who abhorred waste, Corfield had no time for lame ducks. His attitude was to 'stop doing things that lose money and put your resources into those that are making money. It's like taking a weight off one side of a scale and putting it on the other.' Of his own accord he promised ITT Europe that the results of his strategy would be evident in two years. He did not have a master plan but the will to deal with the good and the bad, once they were identified.

His first action was to strengthen the strong. A technology in which STC had been conspicuously successful was undersea cables, its most profitable activity. When satellites became commercially operational from 1965 there was a hiccup in the business because some cable systems were postponed in expectation of space marvels. In a sober assessment of the new situation it became clear that the two media would coexist, satellites being cheaper where random access was wanted and cables offering economy and quality on point-to-point links. The apparently competing manufacturers did not want their medium to be left behind by the other so technological progress was assured and administrations found security in a 'belt and braces' policy.

STC's concern was that its split organisation was not the best way of handling the business. Cable and repeaters, more often than not forming an integrated system, came under different groups. Repeaters, now transistorised, were the responsibility of the transmission group, which had its own relationship with the Post Office, not always to the advantage of the cable people. The Post Office tended to favour its British competitor, Submarine Cables, which in the late 1960s was instrumental in winning some North Sea contracts. STC did what it had often discussed, reorganised its two interests under one management, and then in 1970 it was fit to take over its competitor in the rationalisation following the merger of GEC, AEI and English Electric. In doing so it acquired a company with roots going back to the very beginning of international telegraphs in the mid-nineteenth century and also became the sole national manufacturer of undersea cable systems.

During the 1970s, a decade in which its parent ITT got into and out of the undersea cable business in its own right in the USA, STC continued to score successes. The record was impressive enough for Behn to have given two cheers from the grave. In the Mediterranean there was a criss-cross of cables linking Africa and the Near East to Europe, meeting the needs of trade and tourism. Businessmen used the lines by day and in the evenings tourists queued to make calls home. In 1971 the world's first 1,840-circuit system went into operation between the Spanish mainland and the Canaries; in 1974 the first transatlantic 1,840-circuit system, between the UK and Canada; in 1977 the world's first 3,600-circuit system. Installed between Rome and Palermo, this could transmit a single colour TV channel in either direction simultaneously with 1,800 speech circuits. Increasing capacity continually improved the cost effectiveness of a single cable. That year, 1977, in which the company's largest ever export order to date was booked, for the £40 million Columbus cable between Venezuela and the Canaries, it won another Queen's Award for Export Achievement. With 97 per cent of its production for international markets, it was the world's leading supplier, beating the Americans, French and Japanese. At the end of the year a 5,520-circuit system went into operation on the Spain–Canaries route, to be followed by orders for similar systems linking the UK with Spain and the Netherlands. Undersea cables went a long way to making STC Britain's leading exporter of telecommunications equipment.

In the early 1970s Corfield's urgent priority was to stop the losses, which meant taking rapid and decisive action, unpopular though it might be. In making the company leaner and fitter he had no sentimental attachment to units, young or old, that did not fit into his emerging design or were commercially unsound. Why, he argued, should a firm of telecommunications engineers be making cable drums? We are not carpenters any more than we are cobblers and we are not investing in wood-working machinery. Let us stick to our last. On this principle he dispensed with the company Dakota, in-house printing, translation and the wood shop at Southgate. He got out of marginal activities like closed-circuit television, which the company had been in for only three years. Similarly, he disbanded the project and field services division, another organisation handling a motley collection of assignments that nobody else wanted and that were of dubious profitability. He got his staff to look at things more critically. Do we need it? Is it cheaper to make or buy?

Age and experience in a product were no commitment to its future. Microwave, in which the company had been a pioneer, was chopped in 1971 because it had become a liability. It was a product that required a high engineering investment in relation to sales and these had fallen off in the UK, where most of the mainline network was complete and TV signals were also being transmitted on higher capacity coaxial systems. With the home base-load reduced, it was impossible to devote the amount of engineering manpower and facilities that microwave needed to ensure a profitable future. Immediately, there were also problems in the production of precision components at the troubled East Kilbride plant and losses on work overseas. The company had to negotiate its way out of contracts like the £15 million one in Brazil, paying compensation to customers and profiting from its mistakes by strengthening its contract administration procedures and making microwave a case history on its management courses.

An even bigger case history of how not to do it was the slide from the pinnacles of achievement of the other great pioneering division, radio. In its support it could be argued that it was no longer operating in favourable market conditions. Satellites and undersea cables together publicly spelled the end of high-frequency radio as a means of long-distance communication. The airborne equipment market had shrunk with the decline of the British and European aircraft industries, military business was running down, and sales of integrated ground navigation systems to national government bodies were impaired by ITT's decision to spread this technology among three of its European manufacturing companies. With hindsight, these look more like excuses than problems that could have been tackled. STC had compounded its problems in what was no longer a seller's market by a catalogue of bad management.

Charles Strong, who left the division in 1962 after almost forty years' service, had been a manager who took a detailed interest in the technical aspects of almost every project and had kept the business and the organisation within his personal compass. That was not the style of the ensuing regime aggressively bent on growth. Orders were taken for equipment not yet developed and development had to be completed on sites all over the world at substantial extra cost. Money and reputation began to slip away. To compensate for time lost in sorting out these orders new business was taken, again for equipment not fully developed. The West German defence equipment, a contract so big that it swamped all normal business, was subject to hundreds of design changes, which resulted in production problems and inventory build-up. There had too always been a large manual records section supporting the equipment in service.

This was at a time when the switching division was having difficulties with crossbar and it was expected that radio could make up for those deficiencies as well as its own. Further technical leaps were made into the uncertainty of new categories of instrument landing systems and a radio altimeter, which delayed work on equipments for which there were firm orders. Key staff transferred to the promising new mobile radio division. As part of the downward spiral serious defects were revealed in the standard costing procedures and divisional pricing policy. Splitting into two product lines, communications and aviation, had worsened the situation by creating a top-heavy management structure. Between 1962 and 1971 the division had seven general managers, one of whom cut out all development to improve profitability and another who decreed an end to notification of engineering changes; aviation had four managers in five years, and five comptrollers came and went in three years.

From 1968–70 the division's losses were never less than £1 million a year. In 1969 a team was put in to try and revive the ailing organisation – Frankie Howerd raised the flagging morale of the workforce in the canteen one evening – but things went from bad to worse. Corfield made the decision to phase it out. Bob Coles was the marketing manager at the time:

> As we ran it down I became aware that there were some products that were good and we could have continued to make a profit on. It was impossible for management to sort good from bad while it was an on-going business. They didn't even manage to allocate overheads for example *pro-rata* to the appropriate products. There was a complex mix of products. How do you apportion the costs of litigation on the radio altimeter?

Some bits like the crane control and the altimeters, largely on a care and maintenance basis, were sold to employees. Customers like the BBC were licensed to manufacture or have manufactured products they had in use so that they would be assured of continuity in spares or if necessary complete equipments. Being paid to be relieved of responsibility was an unexpected source of income to the company. Instrument landing systems, considered to be the most salvageable item, were sold to Plessey, which was able to charge more realistic prices for a proven product then much in demand.

What seemed to be a promising survivor was mobile radio, which had started with apparently so much going for it. Technically it was a good product but the management of it was poor, a case of applying muddled capital methods to what was in many ways a consumer durable. In 1970, when the building of a new factory at Radlett, Herts, had been announced, Chris Foulkes was shifted late in his career to try and sort out the division:

> It was split four ways. The accounts were at Rickmansworth, manufacturing at Croydon and Southgate, and they had started on this factory at Radlett. I drew up a plan and asked 'What is the business?' The business is not making mobile radio. Our business is mobile radio engineering. I can get these things made in any television factory. Why are we carrying all this investment, all this manufacturing, when the real secret is the engineering salesmen who sell the type of product, who go out and survey the sites?
>
> We designed a very good product but we were selling it far too cheaply. It had to be type-approved to international specifications, to government specification, you name it. We were not producing to the same standards as television and transistor radios but we were not selling the product for any more money. Pye had 70 per cent of the business and when we started off we got about 8 per cent. At this level you are very vulnerable. Pye had their repair depots all over the country and if anything went wrong they could get the stuff repaired quickly, whereas when our stuff had a fault it had to come back to London. Our people were really living it up. Everybody seemed to have a company car. Thirty per cent of our expenses were marketing.
>
> I drew up a plan whereby we would get the stuff made outside to our designs and have our quality control over it. Then I had to start fighting off ITT Europe, which was holding up as a shining example the mobile radio in Germany. It turned out that the only reason they were profitable was military business. Their domestic stuff was losing money.
>
> I reduced our staff from 700 to 400 and achieved a higher turnover with them than I did with 700, but because we were losing so much money we had no new models. We got up to about 11 per cent of the market and I said: 'The only way now is to invest a lot more money, time and effort. As I see it, it will take us five years to break even. It's going to cost you $10 million. Is it worth it in effort, management? There's no point in having people spending all their time losing money, working at losing money.'

So the company decided to get out and sold it to Pye in 1973. In 1976, after over fifty years in the business, it also got out of public address.

Deciding to cut out divisions and services was something that Corfield could do largely by himself, personally enduring the stomach-turning, especially when some good businesses were removed with the bad. Rebuilding the organisation for a healthy, prosperous future demanded the considerable extra effort of a new top management team, a group of people with a commitment to change as a way of life. Basic technological changes were already becoming evident and people were needed who could prepare for and handle the management of these changes as they affected all functions. Neville Cooper, who came in as administration director, spent two years establishing the structure for the future and getting the team together, a task substantially complete by early 1974. Largely the talent came from outside, a cause of some confusion and dissatisfaction to insiders, who saw colleagues being laid off in closures at the same time as men of ability were being sought. Initially the emphasis was on bringing power back to the centre, the only place from which the build-up of the company could effectively come. Later, as paths of development and policies became clearer, this gradually evolved into a better balance between central functions and line responsibilities for the various businesses.

This philosophy of centralised direction and decentralised management could be seen at work in electronic components, legally part of STC but since 1968 part of an international product group, where internal growth and diversification had continued. It was Britain's largest manufacturer of capacitors, quartz crystals, reed switch inserts and thermistors. Far from being moribund, the valve division had expanded its range of electron tubes, microwave and electro-optic devices. Film circuits were being applied in items like heart pacemakers, standard power supplies in all sorts of applications as electronics found their way more and more into the home, industry and commerce. To meet the demands from so many quarters, distribution organisations were increased and expanded, including taking on distribution of competitive products. In 1976 the group strengthened its overall position in the market, where it was principally known as a supplier of professional components, by the acquisition of Erie Electronics, which manufactured capacitors, film circuits, resistors, potentiometers, and other products mainly for the entertainment or consumer sector. Only the second acquisition in the history of the group, it marked a major stage in its evolution from an in-house source of telecommunications components to an international supplier of a broader range of devices.

In STC the largest single activity was the manufacture of telephone exchanges, on the success of which the fortunes of the company depended. Had it been simply a matter of doing better what was done before, life would have been fairly easy, but the technology was changing from basically electro-mechanical construction and operation to an increasingly electronic form. Although the introduction of crossbar as a stop-gap had been an unhappy experience for the Post Office, it was still placing orders in 1971. At the same time it was also ordering for installations up to 7,000 lines the TXE2 exchange, electronic in that control was by computer. Switching was electromagnetic, by reed relay, and conversations were still in analogue form, electric currents that mirrored vibrations in the human voice. TXE2, a collaborative development of the manufacturers and the Post Office during the 1960s, thus formed a convenient add-on to the existing network but development was needed to produce a larger exchange in the same family to handle up to some 40,000 subscribers. TXE3 was a technically but not economically viable prototype. STC, fully aware of ITT's interest in selling a European system to the UK, knew that the Post Office wanted a domestic product, one that it had a hand in. Ken Frost was then responsible for marketing strategy:

> We took a chance on going into TXE4. In the first place without any firm commitment from the Post Office. And then we got into the great modernisation programme, in which we argued for modernisation with TXE4 while Plessey and GEC made their case for crossbar with some electronic control on it. The Post Office decision was for TXE4.

That put the company in an advantageous marketing position but created so many internal problems that it was in danger of losing the initiative. TXE4 meant large-volume business even though the individual types of racks, shelf and printed circuit boards of which it was composed were not in themselves large-volume items. Together, they amounted to a complete change, not just in technology, as John Smith saw when he moved in the summer of 1974 from the transmission division at Basildon to Southgate, where electronic switching was starting:

> The site was the end of 50 years of growth and development in its prime product and electromechanical attitudes were ingrained at all levels. Many people could not accept the idea that what had existed for so long was vulnerable and could be challenged. There was an apprehension about electronic switching, this struggling infant that was getting out of control yet was obviously going to be the future. A lot of people had no understanding or skill in it. It threatened them and so they resented it and tried to fight against it. It meant a totally different type of labour force, with an increased ratio of women to men, using new mixes of skills, production and testing methods. Machine shops were becoming a thing of the past and printed circuit boards the building blocks of the future. Electronic switching was a different culture.

His brief was to change the attitudes at Southgate and he was soon resented. Some managers were not speaking to one another. Rivalry existed between Southgate and the Northern Ireland plants, with both sides anxious to protect their futures and make sure they did not fail during the awkward process of transition. There was a large and urgent learning programme to go through in the running down of the old and the start-up of the new for a demanding customer, also feeling his way in electronics. As part of its change of direction and to meet the needs of its major product divisions as well as the growing international market, STC acquired in 1974 the UK's third largest printed circuit board manufacturer, Exacta Circuits based in Galashiels and Selkirk in the Scottish Borders. The company, founded in 1962 by two men, had specialised in the professional market but, even so, it was not up to making the new boards without considerable investment. Under its new management, with more capital and the development of its French subsidiary, it was destined to grow twelve-fold in six years.

After six months at Southgate, John Smith almost felt inclined to blow it up and start again. Attitudes were slow to change and dialogues between different groups were only gradually opening up. It was the shock effects of an external event that brought home to people the magnitude of the change and the speed with which it was going to have to happen. The programme of adjustment was under way when in 1975, without warning, the Post Office brought down its axe. For the third year running its telecommunications activities were in deficit, this time with a record loss of almost £200 million. To help make a quick return to profit it cut back sharply on its electromechanical ordering programme. Some idea of the sharpness of the change can be gathered from a glance at the company's ratio of electromechanical to electronic exchange sales: in 1974, 86 to 14; in 1978, 34 to 66. The first casualty was the Larne plant in Northern Ireland, which was run down and closed in 1976. There were also redundancies in Monkstown and Enniskillen, and at Southgate. With the added closure of East Kilbride, the company altogether reduced its workforce in this product line by about a third, from 21,000 to 14,000 people in two years. The last Strowger rack was made at Southgate in 1979.

Back in 1975, when the company was coming to terms with this sudden shift in its main product, nobody could spell out the exact consequences but the writing was clearly on the wall. At Southgate some of the traditional types moved over to electronic switching and gave greater credibility to the new enterprise. Success helped breed a new team spirit.

Early in 1976, the centenary year of the telephone and the year in which the Post Office's last manual exchange was closed, the first TXE4 went into public service, at Sutton Coldfield. It had only 5,000 lines but it quickly justified the risk that STC had taken. The Post Office, relishing competition, issued a bullish statement on its significance and made STC the major supplier, with the other two companies having to catch up. Sales and profitability started growing. Two years later Ken Corfield was able to say at his management conference:

> Under the old bulk supply agreement, STC had only 20 per cent of the main exchange business against the 80 per cent that GEC and Plessey shared. Our present participation in this business is at the 35 per cent level – surpassing either competitor. This entirely reflects our own initiative. It was STC that took on the TXE4 development contract when the others refused it. It was STC that developed TXE4A to guarantee a 15 per cent reduction in costs on TXE4 while providing very much better facilities. Our present 35 per cent participation will probably rise, but the Post Office has indicated that STC can expect, on a continuing basis, to be regarded as at parity with the other competitors. This is not unfair, even though it underestimates STC's competitive edge.

A factor in the design of TXE4A was interworking with the fully electronic exchanges that were only a few years away. In these exchanges the operation would be fully digital, with conversations represented as a series of digits or bleeps, like information in a computer. Since mid-1971 an experimental STC exchange working on these principles had been in public service at Moorgate in the City. Developed in association with ITT companies in Europe, it was a pointer to the way things were going. Computers and communications were becoming integrated. A system was no more just a piece of hardware but had a growing intellectual content of software, which gave it flexibility and enabled it to perform so many more functions. Whereas electromechanical exchanges were comparatively bulky, labour-intensive to produce and maintain, electronic equipment was smaller, more reliable, and could be produced and tested by more automated methods. As more complete and tested units were shipped, fewer installation staff were needed for site work. New technologies demanded new skills and people had to be retrained.

To create work and to fill some of the gaps in the services provided by existing electromechanical exchanges, the switching division devised a number of new products: a device for making an operator's job easier and more productive; a cordless switchboard; a push button version of its warbling telephone, the Quickstep Deltaphone; a teleconference system enabling people to meet without travelling; and a test desk for detecting and identifying faults in subscriber lines and telephone sets. At Newport the electronics division had been moved from the cable site and, master in its own house, was competing in the market for special accessories. There were other reshufflings of manufacturing to improve operating ratios but, worthy though these efforts were, they could be no match for the new situation. The effects of the recession following the oil price increase were exacerbated by the process of fundamental technical changes which, even when markets expanded again, would mean fewer jobs. In the mid-1970s apprentice intake was slowed down. Altogether rationalisation costs between 1976 and 1978 were put at £20 million, with a further £6 million in capital expenditure.

Part of that cost was for transmission division, where similar changes had been taking place, as Bernie Mills outlined to the management information meeting in 1978:

> In the Post Office transmission market we have had to cope with a 40 per cent decline over a five-year period. This market downturn has been caused by the economic recession, tariff changes introduced by the Post Office to spread the busy hour throughout the day, and the Post Office's asset-utilization programme, based on the use of computers to get more effective loading of the network. The result of this market collapse has been a serious over-capacity in the UK for the last few years, with poor price progress. We are now in the trough of this market, but for the first time have some confidence that the PO market will recover.
>
> Our problems of a collapsing home market have been compounded by a major change in the fundamental technology of transmission. The UK telephone network is an analogue system with space switching, but in the 1980s this will change to a digital network with digital switching when System X exchanges are installed. To prepare for this digital era, and to obtain the full economies of this transition, the Post Office are progressively changing the transmission network from analogue to digital – from frequency-division to time-division multiplex – over the period 1976 to 1989.

The transmission division was technically well placed for this new era, having installed Europe's first high bit rate digital transmission system. Operating on an existing coaxial line between Guildford and Portsmouth, the system had a capacity of 120 million binary digits per second, the equivalent of 1,680 telephone conversations, a colour TV channel, eighteen viewphone channels or up to 224 music or broadcast speech channels. It could also claim the first 140 million bit system, suitable for optical fibres; the first thirty-channel pulse code modulation systems in the UK, in regular service forming the first level in the planned digital communications network; and a Post Office study contract for a digital line system capable of carrying nearly 8,000 simultaneous trunk telephone conversations. In addition, the division had also maintained the development of its coaxial line systems for export, for example to Egypt, Malaysia, and South Africa, where after sales of interests in the latter part of the decade the only remaining local holding of STC was 30 per cent of the cable company ATC Ltd.

Within STC these technical developments resulted in an important physical change. Transmission and switching were both now based upon printed circuit boards, which had to be produced competitively. It therefore made technical and economic sense to develop a facilities plan for the two products, to merge their manufacturing under one roof, where what had been separate functions could now both benefit from a single capital investment in facilities like automatic stores for component provision, lines for flow soldering of boards, and automatic test equipment. The decision was made and announced at the end of 1977 to concentrate manufacturing at Southgate, a move that would take two years to accomplish. Not unnaturally, many employees who only fourteen years before had transferred from Woolwich to Basildon felt that the integration should take place there in the newer factory. This was taken over by the consumer products group, which closed its older, smaller plants and concentrated its activities there.

It was not out of bloody-mindedness that people did not move from Basildon to Southgate. There was an enormous cost difference between housing in the two areas. In Basildon houses were either rented or, if they were owned, would have fetched far less than the cost of buying property in a London suburb. Only seventeen employees moved and, although there was a temporary travelling scheme between the two sites, it took much longer than planned to re-create on a different site a purpose-built transmission factory with all its skills. Deliveries fell behind, customer confidence and market share fell. For a while it looked as though another major line of products in which the company had enjoyed a technical lead and a world-wide reputation might be in jeopardy.

Nor was the traditional cable business immune to change. Since the end of the telephone cables bulk supply agreement in 1963, to protect themselves the manufacturers had entered into unofficial arrangements that more or less perpetuated the old system. These arrangements came to light during an investigation by the Monopolies and Mergers Commission, resulting at home in a lull in orders and a competitive tendering procedure. This all happened in the difficult years following the sharp increase in oil prices at the end of 1973, when there was a world-wide over capacity in telecommunications cables and aggressive competition in exports from Japan, Korea and Taiwan. Throughout 1975 doubts about the future were growing among Woolwich employees, who saw repeater manufacture moved across the river to Greenwich. Ten days before Christmas the general manager, Peter Thew, who had first come there as a school-leaver, told them that the plant would be progressively run down from mid-1976 and closed by the end of 1977. Production would be concentrated at the more modern factory at Newport, which had the advantage of being single-storey and which had gone over from paper to plastic insulation, a development partly made necessary by the change-over from copper to aluminium conductors.

The statement, which gave the employees a rotten Christmas, was like telling a Catholic that Rome was no longer the home of the church. Woolwich had been the foundation of the company. The sweat and toil of generations were part of its fabric and it was unthinkable that all that should come to naught. STC (Stop Thew's Closure) stickers appeared and effigies were burned. Some of the 1,200 employees, not just on the shop floor, could not accept it and would not talk about the unthinkable. There were those who believed it hastened the death of the 88-year-old Sir Thomas Spencer two months later. One bright shop-floor idea was to sell the plant as a going concern to the Saudis, who had plenty of oil money and needed telecommunications. But the plant was not a commercial proposition, a situation eventually accepted, and socially the community was not what it had been. The neighbourliness of the terraced streets had vanished, replaced by the isolation of high-rise blocks, the inhabitants of which jumped in their cars and drove to high-paying employers like Ford at Dagenham. There was a high labour turnover. Although the area was running down because of the move of the docks down river and the closure of other industries, there was nevertheless a strong emotional attachment to it. For those who moved from Woolwich to Newport it was like 'going to a foreign country'. The plant was sold to a property developer and in 1981 some of the buildings were demolished.

Also produced at Newport were special-purpose cables for which a considerable future was seen: the electrical and electromechanical types that provided power, communication, control and monitoring links, originally in defence applications like torpedo guidance but now, with the exploration and exploitation of oil and gas in the North Sea, increasingly for civil use as well. They might provide lifelines to divers, carry control signals to undersea wellheads, or a variety of services to unmanned plant engaged in deep-sea mining. Altogether some 500 different types, along with accessories and mechanical handling equipment, had been developed when the hydrospace division was formed at the start of 1976. Among the kaleidoscopic beauty in cross-section of this growing range of cables one of the more easily understood products was a plug and socket enabling a live connection to be made underwater.

As Woolwich started to run down, production of a new type of cable started in Harlow. Ten years before, in 1966, two STL scientists, Drs Charles Kao and George Hockham, proposed a glass fibre cable carrying signals on beams of light through tubes as fine as human hair. Vastly more information could be carried over an optical fibre cable than over a conventional cable of comparable size. Cable of this kind could often be installed in an underground duct too full to take any more conventional cables and its size and light weight made installation faster and cheaper. Signal losses were lower so longer lengths could be used and fewer joints were needed, reducing the need for intermediate repeaters. It opened up the prospect of an integrated broadband network carrying economically speech, data and moving pictures. Glass fibres, rugged and flexible and, unlike metal conductors, not subject to electrical interference, promised to be the highways of an information-based society, the high-speed links for electronic mail, cable TV and other services. The significance of the development, which would not become generally apparent until the early 1980s, was appreciated by the technical director, 'Jock' Marsh, who allocated money and resources to encourage it.

In 1976, as a private venture undertaken by STC with Post Office co-operation, STC and STL installed a field trial system over the six-mile route between Hitchin and Stevenage in Hertfordshire. Able to handle 1,920 simultaneous telephone conversations, it was reckoned to be the world's first high-speed, high-capacity optical transmission link using a repeater. After fifteen months' continuous operation, during which the BBC also successfully conducted a series of colour TV test transmissions, the link carried normal public telephone traffic. In the meantime the company had been asked by the Ministry of Defence to develop a similar system for military application. This was followed by a contract, more Post Office orders, and growing interest from overseas. These were only the beginning of the exploitation of the new technology. There were many other possibilities, among them application in undersea cables, where the pay-off in vastly increasing the capacity of a single cable could be dramatic.

In the midst of fundamental technical change the company had to work out a *modus vivendi* with the Post Office, which was finding its new role as a public corporation dealing with suppliers no longer in bulk supply agreements. In telephone exchange business STC had benefited, its faith in TXE4 enabling it to establish what was to be an enduring parity with the other two major suppliers. All the manufacturers though were under something of a cloud while Sir William Ryland was chairman of the Post Office. His working life had been spent with the organisation, he got involved in a mass of administrative detail and he was suspicious of the manufacturers being in cahoots to keep up price levels. At TEMA annual dinners he remarked that profits on Post Office business were never really disclosed and were hidden elsewhere. Pressure was put on prices.

The transmission people in STC had seen tenders awarded on price alone, irrespective of considerations of quality, technological advance and research expenditure. In switching there were cost investigations after contracts had been completed, even though the quotations had been subject to an agreed schedule that had been cost investigated. To remind suppliers that they were not in a captive market, the Post Office made some purchases from abroad and threatened to increase them. In its defence it could argue that in cables its policy had been 'buy British' but that the industry had let it down by operating an unregistered agreement that fell foul of the 1956 Restrictive Trade Practices Act. Detailed investigations were carried out by the Post Office and chartered accountants Coopers & Lybrand for the twelve years that the companies were operating their private agreement. As a result, in mid-1978 the four cablemakers – BICC, Pirelli General, STC and Telephone Cables – agreed to repay the Post Office a total of £9 million by way of adjustment to the prices charged during the period before 1975. On business worth £450 million it amounted to 2 per cent of the value. More important, a point of principle had been established: the end of the old bulk supply arrangements. Although he retired in 1977, it also in a sense marked the end of Sir William Ryland's chairmanship.

The negotiations on the terms of the settlement were conducted by his successor, Sir William Barlow, with the chairmen and chief executives of the companies. Coming from private industry, he had a better understanding of the problems they faced. He was going to need it because both parties were moving into the infinitely complex world of fully electronic telephone exchanges. The mysteriously named System X, after thirty years of experiment, failure and limited success in electronic switching was going to demand a new relationship between the Post Office and the suppliers if it was to be a national achievement. Post Office specifications had been less and less helpful to manufacturers seeking business in export markets and already some overseas competitors were claiming a lead in the new technology that was going to be at the heart of telecommunications developments in the 1980s and beyond.

15
Telling it Like it Is: The 1970s

In 1973 Anthony Sampson's 'secret history' of ITT, *The Sovereign State*, was published. The book portrayed a conglomerate chameleon-like in character and imbued with the low morality of the Nixon administration at the time of the Watergate scandal. It was a portrait half-heartedly defended by the European companies, which in their own countries adopted the attitude of Southern gentlemen in the American Civil War, referring to the Yankees as 'those people over there'. If ITT was a sovereign state, then STC was a very different polity, more like a Greek city state. It faced problems that in miniature were those of Britain. The company was having to adapt to change, not just the fundamental shift in its technology but also the more competitive conditions of freer trade within the European Economic Community, which Britain joined on 1 January 1973. At last it was having to come to terms with its post-imperial role.

In reshaping the company hard decisions had to be taken, which as Corfield pointed out to his people:

> were not in any way dictated by ITT. All the things that we decided not to do are still being done in some parts or other of ITT Europe where conditions and resources and economics and tradition and customer needs dictate that they should be done or enable them to be done with reasonable economy. No, they were not dictated by ITT; they resulted entirely from our own management assessment of the situation.

In fact, the decisions were supported by ITT.

Strategist though he was, the direction in which he was leading the company was not immediately apparent to everybody. Initially, he appeared more as a demolition contractor than a builder and when he did build he brought in specialists from outside, an action hardly likely to raise the depressed morale of those who felt they had given good and faithful service and were being passed over. He extolled the principle of change with a Darwinian enthusiasm but he recognised that not everybody could be expected to see the evolution of the company from his viewpoint. To the individual affected by new market and technological forces, change could mean a bewildering personal upheaval.

When Corfield had got the company moving in the right direction he therefore sprang on his management board, which would have appreciated a respite to consolidate the work in progress, a new challenge: 'We should aim to be the best company to work for in this country'. His proposal was greeted with a sceptical silence and then questioned as something that might be conceited, undesirable, unattainable and impossible to measure. After some discussion, the board came round to the view that it was better to have something than nothing to aim at, to miss rather than not try at all. The main responsibility for spelling out the policy belonged to Neville Cooper:

> Exactly what the aims were had to be worked out, not as a set of paternalist principles imposed from above but as an expression of what people at all levels in the company felt. We asked them what they saw as features of the best company to work for. Finding out that it was important to employees that the company gave good service to customers and was an effective and profitable organization was the beginning of what became a developing process of consultation, an exchange of facts and feelings on issues that mattered to people. When we'd found out we put the principles in writing.

One of the main principles in being socially conscious on a par with being technically advanced was that the individual should be treated with respect. He had a personal stake in the business and an opinion that was entitled to be heard, in the way that the company expected its own views to be listened to by its parent. During the early 1970s, when ITT was facing a barrage of press allegations about its conduct in Chile and elsewhere, the internal criticisms were just as fierce. Senior company personnel argued to New York that the allegations had to be answered, which New York was going to do anyway. Its stand was subsequently vindicated by the Overseas Private Investment Corporation, settling ITT's $125 million insurance claim for the expropriation of the Chile

Telephone Company, and the testimony of two of its severest public critics that it had done nothing illegal and was in no way involved in the coup. Similarly, STC wanted a cards-on-the-table, man-to-man relationship with its own employees for a constructive dialogue. It was a laudable aim that could easily generate cynicism, but nobody was going to make a stupid claim that it was the best company to work for. Rather was it an ideal to aim at and the final judgement could only come from the employees themselves, whether they felt they would sooner work for STC or another company. There was much going on, large and small, to put the philosophy to the test in a rapidly changing environment.

Like a successful semiconductor company, STC was effectively moving into smaller premises. It was becoming more capital- and less labour-intensive. As the decade progressed Post Office inspectors, who even after the ending of the bulk supply agreements still kept their eyes on costs and quality, saw the machine shops and the intricate craft of the wiremen giving way to lighter, more automatic methods. There were fewer people on the shop floor, casual in jeans and with cleaner hands. Yet the apparent relaxation did not mean that less work was being done. In ten years there was an eight-fold increase in capital investment per employee in manufacturing equipment. Whereas in 1970 manufacturing costs of electro-mechanical equipment were 30 per cent material and 70 per cent labour and overheads, by 1980 with the existing electronic technology those percentages had been more than reversed. The trend would continue.

With computer-aided design, from original logic to final physical layout, weeks of work were being saved on every printed circuit board. The techniques could also be used in making modifications to existing boards, with the added administrative benefit of updating component lists. With computers, design and production were not so much different functions as aspects of one process. Moves towards integrated information systems enabled more economical stock levels to be kept. Process research was also being done at STL, which co-operated with the automation unit and product divisions in STC to produce machines for the precision mass manufacture of different types of electronic components. For equipment manufacture, components called up from automatic stores could be selected by machine and put into the correct sequence for automatic insertion into boards, which were then automatically tested. Numerically-controlled machine tools were being applied, laser machining, and at Paignton the ultimate in programmable equipment, a robot. All this involved less paperwork. Instead, computer terminals with visual displays of columns of figures were in evidence, transferring information much more rapidly to the involved departments and lowering the ratio of indirect to the reduced number of direct employees. In offices word processors were beginning to be installed.

As computer and communications technologies converged, telecommunications manufacturers, who had hitherto enjoyed a fairly stable market for a repetitive product, were finding competition in PABX systems from giants like IBM. As the pace of technological change quickened it became obvious that products would have a shorter life cycle. Their diminishing hardware content was less important than the software, the intelligence that gave them flexibility and enabled them to do more. Often the investment in this element was the greater. Hardware engineers were having to learn software methods and a new category of person was emerging, the software engineer. In multi-disciplinary teams it was those who could adapt most – and these were usually not the programmers, who tended to retreat into their craft – who became the leaders. The mix of required engineering skills was changing and more graduates were needed with electronics and production backgrounds. They had to have human skills too so that they could lead a quality circle, a small group of employees improving processes and relationships in their own area.

A novel effort to boost recruitment in 1975 involved sending a record giving the facts of life in the company to respondents to advertisements in the national, technical and local press. It helped attract people who would otherwise have gone into computers or other apparently more exciting products. Not keen to know were some students at York University, where a resolution was passed that the company should not be allowed to interview on campus 'because of the involvement of its parent in the overthrow of the Allende government in Chile'. Rather than get involved in sorting out allegations to which it was not a party, the company changed the place of interview for the first time in its graduate recruitment campaign.

Overall, the number of people in the company was falling and in a period of change it was important that they should know what was happening to them. Not that all division managers immediately saw it that way. Some resented headquarters going over their heads to their own people. Yet simple downward communication from the few to the many, the paternalism of Spencer (who left £243,934, scarcely a fortune, from his life's work with the company) and Pheazey (£145,454), was no longer adequate or possible. People wanted to know why the partitions were being changed around them before any move was made and, if they were to give of their best, they had to be not just informed but involved, understanding their role in the whole. They came to look upon the company not as something that owed them a living, but a living thing that they helped sustain so that it provided them with a living. A simple example was the change in pension arrangements, with the non-contributory scheme of the cap-doffing era giving way to a contributory system. For a while the two co-existed and then the non-contributory was dropped. When the new provisions were being explained, with the help of audio-visual aids, to several hundred small groups there was some interesting feedback to the management, for example about the working conditions of part-time and female employees.

To ensure that not just lip service was paid to the idea of communicating, the company created the job of manager, environment and communication. The first person in the new seat was Mrs Pat Webster. Her task was to improve the whole working climate, the physical environment and the two-way flow of information. The appointment was enlightened, possibly the first of its kind in the country, yet it did not make the company liberated in the eyes of the women's movement. Later STC could claim that it had a higher proportion of women scientists, technologists and supervisors than other comparable companies, but women had still not risen to director level.

The importance of communications in a manager's function, an essential element in leading and holding the confidence of his staff, was stressed in management training courses. Communications were as important as product knowledge, being able to interpret figures or a critical path analysis or appreciate the significance of microprocessors and the place of programming. Handling them properly contributed to the success of projects, which were now bigger and had to be accomplished more quickly. However complex the technicalities became, the organisation was still people-based and they were the major resource that a manager would have to manage. To do that he had to be willing and able to communicate, not putting off the evil day for fear of reaction but facing the problem and using appropriate techniques to put his message across. There were bound to be circumstances in which he could not delegate the function. He had to have a personal commitment to it. It was an aspect of the new attitude as defined by the industrial relations manager, Jack Skipp:

> We conduct our affairs not on the basis of the right to manage but of the responsibility to manage. Standing on rights leads to stand-up fights. We have to persuade, lead, educate people that change is the right thing to do. There is a middle way between 'may' and 'I will'. The easiest thing in the world is to say 'No'.

On the whole industrial relations were good. In the first half of the 1960s the forty-four hour working week had by national agreement progressively been reduced to forty hours. For that time a higher-grade skilled man was by the mid-1970s earning £100 a week. The drop to thirty-nine hours did not come until 1980. In 1970 three weeks' holiday for manual workers was won and by 1983 this would rise to five weeks in addition to the eight statutory days. Within the company working conditions improved through a combination of pressure from below and a greater willingness from above to spend money. At Southgate, for example, the institutional concrete floors and ceilings and the army-style cream and green paint were softened by false ceilings, diffused lighting and warmer colours. In headquarters a self-service restaurant with a club atmosphere was created for the use of everybody, directors and staff. Working atmospheres improved, as evidenced by a revived interest in the suggestion scheme. Awards around the £2,000 mark were something to shout about in the middle of the decade and by 1979 the highest ever award came to £15,198, for an improved method of inspecting reed relays. It made Mrs Jo Bevan of ITT Components in Harlow feel like a football pool winner. There were a number of multiple winners.

Another interesting figure from the mid-1970s, the period of greatest change, was that industrial stoppages amounted to about three hours per year per employee, including 'nationwide political stoppages'. One of the reasons for this creditable figure was that bargaining continued to be on a plant basis, about local conditions, and plants retained their individual identity. But good labour relations are not just an absence of strikes as unions can have a negative effect by exercising an undue control over production processes. At Southgate up to about 1970 there was an unholy alliance of line management of the old school, engineering- rather than production-minded, and shop stewards with a vested interest in a historical cash piece-work system. It took nine months of negotiation to remove much of the bargaining power of this and establish a new pay deal that changed the basis of industrial relations. In the new atmosphere it was then easier to discuss the introduction of new technology in what had been regarded as 'the graveyard of innovation'.

None of the changes could take place without large redundancies but these were not the cause of industrial disputes. Like the phase-out of the radio division at Southgate and the cable division at Woolwich, they were handled personally according to a detailed plan. The man responsible for the first major phase-out was George Heard, a technical man who had grown up on the Southgate site and knew it and the people well. Round him he gathered a team to take care of all aspects of winding down the business. When he was personally held responsible for the sad state of affairs by a district official of one of the trade unions involved, a senior shop-steward sprang to his defence, threatening to withdraw his delegation if the official continued in that vein. Within the company the union/management relationship was characterised by informality and a willingness to face the facts of a situation. Attempts by outsiders to make political capital out of issues were jointly resisted.

Four years later the principles laid down in the best company philosophy received an early and unexpected test when the Post Office axe fell on electromechanical switching. The company had agreed plans with its employees and the trade unions for the change-over to electronics, in which there would be a steady rundown of the old product and a build-up of the new. As far as possible the aim was for products and people to serve out their time together, with some people being asked to stay on a bit longer, some retiring early, and some being retrained. In the change-over there would be no attempt to try and mix the old and the new in the same premises so that everybody would know what was expected of them. When the axe fell the programme suddenly had to be accelerated by two years. The event cast doubts on the good intentions but the company took the view that, if its social policies meant anything, now was the time to live up to them. People were given the facts as soon as they became available and the best mutually made of a difficult situation. Altogether 7,000 people went, without confrontation.

At Woolwich the key people had been apprentices there and knew many of the workforce by first name. It was reckoned that, but for them, there would have been a sit-in. All the options for the future of the plant were examined, objections to the closure answered to the point where the situation was accepted with understanding. After the negotiations were over the general manager, recovering from an operation, received a get-well card from the woman leading the action committee. There was an instance of somebody driving a nail through a cable but output and quality were maintained and the final party was like a riotous wake.

Perhaps because of this good working atmosphere white-collar unionism never took off to the extent that was predicted, except in the Celtic fringe. In 1970 there was a strike of clerical workers at Newport, an event previously unheard of, which resulted in the salaries of junior staff grades being locally instead of centrally administered. By 1980 less than half the company white-collar staff were bargained for and then by individual unions, which would not sit down together like the manual unions. It was in the traditionally militant areas that class attitudes died hard and disputes occurred. The most troublesome plant was East Kilbride, where a shop-steward with political ambitions tried to make a name for himself by exploiting economic grievances. Some of the troubles were documented in a 1971 report of the Commission on Industrial Relations and in a play by the 7:84 theatre company, which took its name from the statistic of 7 per cent of the population owning 84 per cent of the nation's wealth. The 'them and us' attitude was leavened in Northern Ireland by humour and a loyalty to the company that was providing work. When East Kilbride was on strike work was moved into Northern Ireland, where the shop-stewards objected not to the principle but to the labels of origin. During the national engineering one- and two-day strikes of 1979 the only two places to come out in support were Northern Ireland and Treforest.

By and large the assimilation of immigrants, who formed an increasing proportion of the company's workforce, proceeded smoothly as people of different cultural backgrounds got used to working alongside one another. The company helped integration by providing language classes. There was only one incident that flared up into newspaper headlines. In 1970 at

Southgate the company wanted to promote a West Indian operator, who belonged to the AUEW, to a setter. On a shop-stewards' committee a West Indian made a derogatory remark about the setters – mainly white and members of the ETU – to the effect that they didn't set up machines promptly enough and deprived his members of their bonus. The setters retaliated by refusing to accept the promotion until they got an apology. A shop-steward with a grievance argued that the company should take disciplinary action against the setters, whereupon the setters threatened an all-out strike. The company was caught in a fork. If it disciplined the setters then the whole factory would come out; if it didn't act it faced a racially based strike, which is what happened. Although not fully supported by coloured employees, it attracted the attention of do-gooders and others and involved a picket at STC House. The company, which had a firm policy of non-discrimination, treated the dispute as an industrial matter and sweated it out until the strike collapsed. The operator was promoted.

Like many immigrants he was ambitious, feeling that he had to do that bit better to establish himself. That feeling was becoming more common as people realised that their working lives could not be based on one technology, that they had to be prepared to adapt. It was not always a matter of redundancy because they could retrain and possibly relocate. Company policy was redeployment whenever possible. One of the services born out of the closure of the radio division was Jobscan, a list of junior and middle-grade jobs available within ITT in the UK. Vacancies often ran to a few hundred. Much larger in total were the retraining schemes. For the installation of new types of telephone exchanges some 7,000 people had to be retrained, which involved aptitude testing, courses in basic electronics, classroom and practical experience on the products. In carrying out the programme the company learned that people were more adaptable than it had given them credit for.

The whole process of cultural change involved a lot of effort within the company but the quiet revolution passed almost unnoticed in the outside world, where sit-ins, demonstrations and experiments in workers' co-operatives made the headlines and TV screens. On the evidence provided by the media, organised labour in the 1970s seemed to be resisting change and at best preserving jobs. Yet it was the companies that were adapting that were best fitted for survival and in doing so hundreds of individual dramas were being played out, quietly, painfully, hopefully, bitterly, thankfully – no two were alike. The situation was much more complex than simply getting a job in the 1930s had been. For many employees at all levels it meant coming to terms with themselves, submitting to being assessed, having to go to school again, moving to a new job perhaps in a new community, readjusting status and expectations as one competed with younger people in new skills like programming, even establishing a new identity.

Ultimately people had to stand on their own feet. The company saw its role as handling change in a way that satisfied the needs of the business and was at the same time least damaging to the people concerned, all of whom were individuals. In carrying out that policy at least one personnel manager had the charge levelled at him that 'his department had gone Socialist'. In practice it was not simple welfare, more a matter of informing and listening. Policies on redeployment and retraining were spelled out. Planning for rundown, transfers, amalgamations and the like was even more thorough than for a venture into the unknowns of a new market. Individuals were counselled. Throughout, the emphasis was on the development of employee communications, the creation of an open society in which problems were shared not buried.

Corfield encouraged it from the top by increasing participation in the new style annual management information meeting. He involved more managers lower down the hierarchy, particularly the young and up-and-coming ones. He encouraged questions sent in before the meeting and supplementaries raised during it to be answered by the relevant directors. Questions not dealt with at the meeting for lack of time were answered in the printed report sent to managers soon after the event. Each year had its theme and the facts and figures of life were presented so that nobody should be in any doubt where the company stood, be the situation good or bad, and where it was going. The general idea of the meeting still conformed to the company's initials – strength through communications – but the atmosphere was not of a cheerleading rally. 'Top management must encourage all managers and employees to communicate the true picture – even if it is a difficult one – rather than produce the kind of picture that they think top managers would like', said Corfield at his first meeting. At later meetings, in what *STC News* called the

Great Debate, he asked the 1,000 people present to communicate the essence of what had happened that morning to twenty others not present. In that way the word would be spread throughout the company.

Face-to-face employee communications were developed as the basis of participative management. Consultative committees discussed issues like health and safety, the environment, business prospects, orders received, production schedules and organisation changes. Whereas the principle of a change was generally an imposed decision, the manner of its being carried out was negotiable and room was left for manoeuvre. Briefing groups, small enough to provide the right environment for questions and discussion, went into facts and reasons, providing a direct communication with employees rather than via shop-stewards, who could not be expected to act as management representatives. The two small sites where this form of communication started with encouraging results were Treforest and the electronics division at Newport.

At Treforest a two-year experiment was carried out to see whether people wanted to be involved in running the business. The short answer was 'Yes', but certain aspects of company organisation and attitude were preventing full participation. Effort was put into helping first-line management to adopt a more open style, to establish more two-way communications. A joint management/shop-floor project was embarked upon involving the suggestion scheme and health and safety, both of which to the personnel function seemed fairly routine and not genuine participation. In one year suggestions went up by 400 per cent, accidents down by 75 per cent, and a basis for progress in other areas was established. Employees learned that a prime concern to the management was cost reduction to win more orders so targets were discussed, agreed and in an atmosphere of goodwill exceeded.

In the electronics division at Newport there was an open and informal style of management, with disclosure of information almost 100 per cent. Orders lost or gained were announced over the public address system and employees were told about visitors to the division and why they were coming. Financial information was supplied regularly, in detail, including net profit. Initially the unions were suspicious and brought in an independent accountant to make sure that they were being given sound information.

Neville Cooper assesses progress towards a more equal and open society:

> Loosening the rigidity of hierarchies by opening up lines of communication is a gradual process best carried out in a small environment. Inevitably there has been some falling off but there were also permanent gains. The problem is to apply what has been learned to larger sites, where the returns can be commensurate.

Supplementing face-to-face communication was printed information. Under the editorship of Paul Eddy, who later joined *The Sunday Times,* and then John Eve, the nephew of C. W. Eve, who had started the house magazine in 1928, *STC News* grew as a newspaper with an independent voice. Its statement of editorial policy, jointly signed by the managing director and the editor, included not only the reporting and interpreting of company policies but encouraging a feedback from readers and reporting factually on controversial matters wherever possible. Proofs of articles and employee letters did not have to be approved by management before publication. Letters to be answered were given to the appropriate director, perhaps the managing director, who wrote his own reply. A letter on the closure of the radio division was headed 'How could you Mr Corfield?'. Corfield further demonstrated his belief in industrial communication by becoming president of the British Association of Industrial Editors in 1975. He gave his editor freedom and it worked. Readers came to trust the paper, no longer a glossy production but no-nonsense newsprint, as a mature two-way means of communication. Because it did not shrink from mentioning disputes and redundancies, which could be a lead story, or printing criticisms, its good news was more real, not a sop handed down by management to keep the employees quiet.

After a pilot edition for 1974, circulated for comment, an annual report to employees was produced. It contained an account of how each division had performed, a financial statement, an analysis of the way people were employed, lost time, health and safety, organisation structure, pensions, suggestion scheme news and the like. In 1978 an added value statement was included for the first time, from which employees could see what they had put into the creation of wealth and what they had taken out. Notes on financial and other terms were given to personnel managers so that they could explain items in briefing meetings. The presentation in the reports was factual and graphic, with use of colourful bar and pie charts to bring out points. A guiding principle in writing the reports was not to talk down to the audience or to appear to be educating them. Selecting information to manipulate them was also out because discrepancies between what people experienced in their daily work and what they read would soon become obvious.

The company itself had been producing a formal annual report since 1967, gradually disclosing more about itself. It was all part of its emergence from its closed world and a growing awareness of the various publics with which it was having to deal. To establish its status with relevant opinion leaders it invited them to the annual STC Communications Lecture, inaugurated in 1971. The subject was some aspect of communications and lecturers included Huw Wheldon, then managing director of BBC Television, Richard Marsh, chairman of British Rail, John Garnett, director of the Industrial Society, Sir Derek Ezra, chairman of the National Coal Board, Richard Hoggart, best known for *The Uses of Literacy*, Edward de Bono on lateral thinking, and Edward Heath, the former Prime Minister. Corfield had taken over as chairman of the company. At the same time the board of directors, the longest serving member of which was Lord Glendevon, was strengthened by the addition of the influential scientist Lord Penney and City figures like Sir Kenneth (soon to be Lord) Keith and the Hon David Montagu. It was all a carefully managed prelude to going part public.

16

Into the Digital Decade 1979–1983

When Margaret Thatcher won the 1979 general election on the slogan 'It's time for a change', she became Britain's first woman Prime Minister. She was also unusual in having a scientific background. In the midst of the country's economic troubles there was talk of a second industrial revolution, a revolution based upon that cheap computer the microprocessor. One application of this was in electronic telephone exchanges. As a means of communication they were in order of magnitude different, different in kind and scale as a high-speed car to a horse and cart. Yet in his STC Communications Lecture in 1976 Sir William Ryland, chairman of the Post Office, observed that we have 'a penny farthing system to help us live with one of the most rapidly changing and complex technologies of our time'. He cited an example of more than historical interest to his hosts:

> Western Electric made last year 4½ million lines of electronic switching. And they did it in five factories. We now have in this country a capacity to produce one million lines of electronic switching. And it is dispersed over nine towns – not factories but towns.
>
> So, very broadly, we are now equipped to do about a quarter of what they do. But we need twice the number of locations in which to do it. Hardly the way to keep our unit costs in check.

With companies the size of Western Electric in the USA casting their eyes on world markets, the home team had to find a new way of working together if, as part of Great Britain Limited, they were to be in the big league to survive internationally. Sir William floated his idea:

> In these days when our own mixed economy is being diversified by combinations of public and private organisations – or rather organisations from the public and private sectors – why shouldn't we consider together voluntarily a mini-mixed economy of our own? Something unique to our own circumstances and to our own situation.
>
> For example, is it out of the question for the principal manufacturers to separate their telecommunications entities from the rest of their organizations and, while retaining their interest in them, designate them to enter into a full and free partnership with Post Office telecommunications and with the other telecommunications entities?
>
> A partnership of the public and private sectors. Giving joint direction and joint endeavour to the achievement of joint objectives. Jointly deciding the races in which we should run. Jointly endeavouring to ensure that we do not handicap ourselves out of the running. And giving joint support to see that we win the races we enter. Sharing success. Sharing rewards.

In private, there had already been considerable haggling on how the work was to be shared. Initially the Post Office wanted collaborative development with cost investigation and then competitive procurement, which to the manufacturers was hardly an expansive approach to a major national technology in competition with North America, Europe and Japan. The Post Office modified its stance, wanting competitive procurement and the right to investigate the lowest bidder. In the meantime the other two manufacturers, GEC and Plessey, were not that keen to see a large involvement by STC. Whereas they had originally gained a 40 per cent share of the telephone exchange business by buying other companies they had seen STC, with the Post Office support for TXE4, achieve in a very short time a share of 35 per cent. That initial success of their competitor in electronic switching was something they had to live with but they saw no reason why it should become a permanent share.

For its part, in 1975 STC was heavily committed to the development programme for TXE4A and decided to adopt a low profile on work for System X. It believed it could afford to bide its time, doing a minor amount of work while the other two manufacturers went on developing the small exchanges and until the contracts for the digital concentrator front end and the large local exchange were to be awarded. For their part, GEC and Plessey thought that STC was not totally committed to System X and could always fall back on the competitive product of its parent, the ITT System 12. That view began to change when STC secured its first major development contract in 1978. That same year it proposed a jointly-held company to handle export business, a proposal rejected by the other two manufacturers. The Post Office came back with the idea of a differential equity, also unacceptable to STC, which wanted to confirm its position as a full and equal member of Britain's electronic switching enterprise.

Technical progress was being made – STC and GEC shared the contract for the development of the medium/large local exchange – but, after all the tactical jockeying, the shape of the industry was far from clear. The Labour government commissioned an independent report, which recommended a rationalisation so that there would be in effect a single supplier, something much easier said than done. An alternative scenario, promoted by the National Enterprise Board and the Department of Industry, supported by the Post Office, was two suppliers, GEC and Plessey/STC. But who was to take over whom? In early 1979 the question was very real but once more it was relegated with the onset of a general election.

As during the decision on what type of automatic switching system to adopt in the early 1920s, there was also the ambiguous position of the foreign-owned company. Where did its loyalty lie? Wearing a bowler, was STC to appear a British partner in System X and, under a ten-gallon hat, a competitor company offering the ITT System 12? A clear-cut assurance was given personally to the Department of Industry and the Post Office by Lyman Hamilton, Geneen's successor as chief executive of ITT, that STC would take part in the whole System X programme independently of ITT. This fitted in with the idea, long mooted, of floating part of STC on the London stock market. It would be of political benefit in emphasising the company's British identity in a crucial situation and was a move that was inevitable at some time against a background of strong national sentiments in the leading European countries. A local shareholding of 14 per cent had been introduced in ITT's German telecommunications company, SEL, in 1977, and in France a complete company had been relinquished in a government-inspired reorganisation of the industry, leaving the corporation's second main company there in a weakened position.

The policy was spelt out in STC's prospectus for the sale of 15 per cent of its issued share capital in June 1979:

> ITT believes that, where practicable, it is desirable for its

major telecommunications subsidiaries outside the United States serving predominantly local markets to have a degree of local ownership. ITT already has telecommunications subsidiaries in several countries in which there are significant local shareholdings and has decided that a similar policy should be adopted in relation to STC.

The prospectus also put on record the respective positions on electronic switching:

While STC is involved in the development of System X, STC will not pursue work specific to competitive systems being developed elsewhere within the ITT group. ITT fully supports STC's participation in the development and exploitation of System X at home and overseas.

Under the reorganisation, the legal entity of STC embraced the main telecommunications activities in the UK, STL, business communication systems, electronic components and electrical wholesaling. The package was attractive enough for the shares, offered a mite nervously at 160p, to be twelve times oversubscribed. Some 3,300 individuals and institutions were successful, including 1,250 employees. The market opened at about 180p, which provoked further protest from company pensioners, who unlike even new employees were not given any preference in the allotment of shares. In July 1979, when Rand V. Araskog succeeded Lyman Hamilton after only eighteen months as ITT's chief executive, STC's own standing in the City was unaffected. By the end of the year the share price had climbed to 242p and it was to go even higher as electronics provided one of the few bright spots among the prevailing market gloom. Further recognition of the company's new national identity came with the award of a knighthood to Ken Corfield in the 1980 New Year's honours list. That same month Harold S. Geneen was 70 and before his birthday he had stepped down as chairman of ITT. As his influence had waned so the sprawling corporation had begun to slim some of the excess it had acquired in its aggressive years, giving itself a more purposeful corporate profile. Having suffered from being regarded as a big bad multinational, it was also showing more sensitivity towards national aspirations.

Because of STC's involvement in System X and the associated Post Office funding of £150 million towards its development, ITT waived from 1979 a portion of the research and development contribution that would otherwise have been payable by the company. STC was established as a collaborator with GEC, Plessey and the Post Office in the creation of Britain's digital switching system, which made its debut at Telecom '79, the world telecommunications exhibition and conference in Geneva. Visitors were able to see on a working local exchange the capabilities of the computer-controlled system, its quiet microelectronics handling at high speed digital pulses representing speech, data or graphic information. The presentation and marketing of System X overseas was an agreed national effort carried out by British Telecommunications Systems Ltd, in which the four partners shared costs equally. This company would not itself tender and enter into contracts, responsibilities that would be undertaken by the individual manufacturing companies.

A year after the Geneva demonstration the first System X exchange, the overall responsibility of Plessey, was inaugurated in the City. Within a fortnight of the equipment coming through the door it was switching calls, on 1 July 1980, six months ahead of schedule and eighteen months before the original expectation. In operation it was twenty times more reliable than existing equipment and it was prophesied that within twelve years the entire trunk network in the UK would consist of digital switching. This would occupy one thousandth of the volume of a Strowger exchange and by 1992 nobody would be connected to one of those bulky relics. As the system spread, people would get the benefits of facilities like abbreviated dialling, automatic alarm calls, call diversion to another number, an indication that there was another call waiting, and itemised billing. More facilities would become available, not so much through changes in hardware modules but from additional software, the total creation of which was being envisaged in man-centuries. All this rather overshadowed the cutting into service in the autumn of 1980 of the first TXE4A exchange, installed by STC at Belgrave in Leicestershire.

On its home ground the Post Office was having its role as a forward-looking business more sharply defined. In September 1979 the government announced what many had long desired, the splitting of the Post Office into two parts, posts and telecommunications. They were two entirely different businesses, one labour- and the other capital-intensive. Divorce, as the Telephone Development Association had petitioned fifty years before, would be to the benefit of both parties so that they were free to go their own ways. They had vastly different rates of growth, characteristics and future. It was not difficult to imagine electronic mail and viewdata making the postman obsolescent by the end of the century. British Telecom, which failed to attract Sir Kenneth Corfield as its first chairman, vigorously continued promoting these and other services. In the wake of the promotional canary Buzby, one of them was a special range of telephones, of which STC made three: the Deltaphone, a candlestick, and another elegant return to the past, the Classic, in a semblance of gold and marble.

For the company it was a new era of creativity, coming up with a variety of products for an information-based world, an economy employing more in services than in manufacturing. Since 1975 Jeoff Samson, the director of telecommunications, had been using some of the profits on staple products to fund ventures into items that he recognised would be required in a society where communications devices, largely based on silicon chips, would become consumer items. It was a development that helped to bring closer together the product divisions and the expanded research labs, a move welcomed by Desmond Ridler, one of the two directors of STL:

> There was an STL policy not to listen to what the manufacturing company wanted because it was unnecessarily constraining. I don't think we had a fruitful relationship with STC probably until we got to the Seventies. We are not here undertaking fundamental research or even fundamental research with some idea of what the application is going to be in time. This is an applied research lab and what we're involved in is the marriage of advanced technology, which is for STL I believe to supply, and the market requirement, which is for STC to say what is needed. A successful product really comes out of a dialogue on what advanced technology can provide and what the product and market requirements are. The development that we have to concentrate on now is the process of technology transfer to STC. Perhaps the biggest problem we have nationally is to bridge the gap between research and development and the actual exploitation of the product in the market place.

In STC organisations were established to identify new markets and develop products for them. Out of their efforts initially came the Novatel viewdata business terminal, a loud-speaking telephone, a radio pager for the pocket, and a transaction terminal for banking and retail use. The first of these terminals was installed at Heathrow Airport for use on currency exchange. STC Electronic Security Systems was created to market systems based upon a computerised central alarm control room developed by the ITT company in Holland. Involving other equipment like night-sight cameras, access controls and intruder detectors, it also ironically got the company back into the closed circuit TV business. Soon afterwards, an information terminals division was formed to bring together activities connected with the office of the future and the potential of telecommunications systems in the home. Between 1950 and 1980 the number of homes with telephones had grown eightfold to 12 million and the predictions were that the distinction between home and office would become blurred as information was distributed to people instead of their travelling to where it was centred.

One of the early lessons learned in this new type of business was that the company would have to be more fleet of foot in getting products on to what would be a very competitive market. The radio pager, not physically a large item, took five years of hesitant development by the company and the labs before a Post Office contract for 30,000 units was awarded in 1980. Growth of this kind of activity also called for a re-examination of the role of ITT Business Systems, which provided PABXs, increasingly electronic, and intercoms for voice communication, record communications equipment like teleprinters and facsimile, and for the expanding data market modems, data terminals, communications controllers, and message switching systems. There were some areas overlapping with STC and some where the organisation was not represented at all, for example word processing.

Moreover the trend was for all this peripheral equipment to become more important commercially than the network itself, rather like the presents on a Christmas tree. Jeoff Samson pointed to it at the end of 1980:

> In five years' time we believe that the network business will be still static in real money terms because physical growth will be off-set by reductions in price due to technology. At the same time the peripherals market we service – the end-user market – will grow to be at least as big as the network.
>
> In 1995 I reckon that the peripherals market – ITT Business Systems' biggest business – will be at least twice the size of

> the network market. All telecomms companies have got to move that way – and I'm talking of the UK market only.

The two businesses had distinct strategies, technicalities, and management styles but the existing organisation was not suited to achieving the best total result in a particular country. ITT Business Systems was an international product group with its headquarters in Brussels, where attention tended to focus on the successful units that would contribute most towards the co-ordinated profit. This was not always in accord with the national companies' interest of offering the range of telecommunications products within their own countries. A small business system, for example, could be classed and marketed by ITT as an item of consumer electronics, a piece of information technology, or a telecommunications product whereas the user saw it as coming from one source in competition with Olivetti, Philips and others.

To tackle that market the organisation was changed. Instead of being answerable to ITT Europe, ITT Business Systems Group in the UK came under STC management from 1 January 1981. This was in line with ITT's new policy of bringing under 'single country' management network and end-user businesses. The policy also had the effect of encouraging *ad hoc* co-operation between countries, for example in the marketing by STC in the UK of microcomputers made by SEL in Germany. Instead of operating under a central management, national companies were getting together informally to maximise their own opportunities in the two types of business. These were very different in character and the differences would become more evident in the UK as the Post Office policy of 'liberalisation' proceeded. In the relaxation of its monopoly other organisations would be able to attach approved devices to its lines and there would be more competition in the supply of such devices to subscribers. At its simplest a new device would be a second telephone of new electronic design, but the telephone was rapidly going to evolve into a miniature terminal with a display and a degree of intelligence. It would be programmed to provide facilities that people had associated with offices but would find useful in the home, like accepting messages. All sorts of possibilities, from cordless telephones to communicating word processors, were becoming a reality. There was scope for ingenuity, which could start with an inventor in his garage and grow into a successful company. Producing telecommunications devices was now very much a matter of light assembly in efficient volume and did not necessarily involve heavy capital expenditure in tooling up.

Jeoff Samson spotted the differences between end-user and network businesses and the effects these would have on STC:

> The nature of markets will change: a variety of purchasers instead of one; high-risk development instead of low-risk and fast and frequent product development cycles. The change in our manufacturing will be from a rather stable business to frequent, short, sharp runs; fluctuating market shares as opposed to a comparatively stable one; and the need for us to supply to a consumer-oriented end-customer and not to have some massive interface that has its own way of doing things.

The precise terms of liberalisation were the cause of considerable concern and much argument. Corfield, a Tory radical before Mrs Thatcher led his party, was almost as great an enthusiast for Free Trade as Kingsbury had been, provided competition was fair and square. In principle STC, which saw itself as the most efficient of the major suppliers, welcomed the idea of making greater use of the basic resource of the network. It saw the development as an opportunity and a challenge at a time when it was being frustrated by the Post Office in getting new technology to the customer. Because it did not want to give business to a single supplier the Post Office had taken seven years to approve the Quickstep Deltaphone, a delay that to the company compared unfavourably with nine years to put a man on the moon. In its evidence to the Post Office Review Committee in 1976 it proposed that subscribers should be given more freedom to choose and to buy or hire their own equipment to connect to the public network. Private contractors should also be allowed to compete with the Post Office in the supply and maintenance of equipment to operate over that network.

When the British Telecommunications Bill was published in 1980 most of the industry criticism, expressed through TEMA, was centred on British Telecom's retention of its monopoly on the supply of PABXs under 100 lines and the maintenance of any PABX on the grounds that it was an integral part of the network. In the manufacturers' view, British Telecom would not be able to service all the makes of exchanges that would come in with liberalisation; there would be divided responsibility when it came to expanding a system; and British Telecom would have an unfair competitive advantage. It could buy cheaper, small exchanges of downgraded specification while demanding over-engineered products from suppliers, who would have difficulty in selling them overseas. After much lobbying, the monopoly on maintenance was relaxed to allow the manufacturers to service digital-stored program PABXs, which amounted to the growth sector of the market. British Telecom's monopoly of subscriber equipment was to be phased out over three years to allow British companies to develop suitable products that would compete with foreign suppliers. In mid-1982 ITT Business Systems introduced its first digital PABX which, with a capacity of 300 lines, was designed for small- and medium-sized organisations.

These moves were all part of the long-term aim of reducing the company's dependence on one customer, which, with liberalisation, competition from the private sector, and government privatisation plans, was changing more rapidly than had been expected. The more STC was master of its own destiny the more it felt it could prosper. Ample evidence of that existed in the continuing success of its undersea cable business: a £23 million contract for the first system to be laid across the Indian Ocean, from Madras to Penang in Malaysia; £30 million for the majority of the cable in the seventh transatlantic system, the world's busiest transoceanic route; a £40 million contract for a system linking Senegal and Brazil; and £20 million for the Greece–Cyprus link. The cable contract of the century was worth £170 million, covering the major part of ANZCAN, over 7,500 nautical miles of cable and 1,000 repeaters to link Canada and Australia with 1,380 circuits by 1984.

It reckoned that it had half of the international market in this business. Prospects were bright too because international traffic had high growth rates of between 15 and 30 per cent a year. Costs of systems to handle these growing volumes of traffic would also come down even more with the application of optical fibres, initially on short- and medium-distance routes. In 1980 the world's first undersea cable with an optical fibre core was laid in the relatively deep tidal salt water of Loch Fyne on the west coast of Scotland. There the performance of the 9·5 kilometre loop could be studied in conditions similar to the North Sea, where if all went well practical systems were to be laid from 1984. The first of them would be cross-Channel, between the UK and France. In prospect was bidding for the eighth transatlantic cable, scheduled to be in service in 1988, its optical fibres carrying as many as 36,000 circuits. Certainly there would be a demand for it in the 21st century as satellite systems used up the radio frequency spectrum.

A technical factor in the company's favour was its lead in monomode optical fibre technology. In 1981 it secured from British Telecom the world's first order for an operational monomode system. This is scheduled to be in service in 1984, along the 27 km route between the UK's first 'wired city', Milton Keynes, and Luton. Across that distance no amplification of signals would be required, which would produce consequent savings in the number of repeaters. These could now be housed in existing buildings on the network instead of in roadside manholes, making access much more convenient. Laboratory tests had also shown that it was possible to carry four times as many phone calls – 8,000 instead of 2,000 – on monomode fibre. An epoch of mass communications was beginning and in the home, for example, the days of a single telephone line and four-channel TV would seem as quaint as the age of crystal radio sets. Further laboratory tests with optical fibre had demonstrated that spans of 100 km could be achieved without amplification. These advances, offering large savings in the cost of transmission, also had enormous implications for the continuing success of undersea cables. In 1981 the marine and cable group enlarged its interests when Ships Radio Services, a Dutch company selling and installing radio equipment and providing radio officers, was acquired to extend the service provided by International Marine Radio Company.

Since the phasing out of the radio division the company had not been very strong in defence work but, following development effort, that changed sharply with the award of a production subcontract from the Ministry of Defence via Plessey, worth some £50 million, on the mobile battlefield digital communications system, Ptarmigan. The production was handled by a newly formed defence systems division, which embraced hydrospace work and a special systems unit handling things like naval telephone systems. There was also research work on electronic warfare at STL, and the Ptarmigan subcontract in a sense involved a partial resurrection of microwave technology because at Paignton, within the components group, the super high-frequency radio relay equipment had been developed. It was work to look forward to at a time when the group, which had further expanded its capacitor business by the acquisition of Daly (Condensers) of Weymouth in 1979, had had to close Rhyl at the end of 1980 and contract at Paignton and subsidiary plants early in the year.

The long-term outlook was for a decline in the number of people employed in components manufacture although the markets in which ITT was strong were growing. In telecommunications growth was steady, with stronger performance in the professional and industrial sectors. Under these headings came military, aerospace, marine electronics, radar, navigation aids, broadcasting and other radio communications, and the world of computers, process control, automation, instrumentation, signalling and security systems, and much more. Far from cutting demand for individual conventional components the microchip – widely regarded as one of the causes of unemployment – was expanding demand, at least until the mid-1980s.

17

Prologue to the Next Century

A business earns its living day by day and a centenary is a working day like any other. The fact that a company has evolved over a century is of itself no guarantee of survival. Like any other creative organisation, it depends on its current strengths. It exists in a competitive, changing world and if it does not adapt it will die.

A glance at the history of STC shows that it has been most successful in the business that it knows and understands: telecommunications. The lesson seems obvious but it has not always been so to those doing the job. Success has come from using experience and exploiting new ideas to stay ahead, whereas diversification has been a straying from the main path.

Yet in the long view, by any standards, STC has been a successful organisation. What began as a two-man one-boy import agency has grown under its two American parents into a high technology business employing just under 22,000 people in the UK with a turnover of over £600 million, and is now a public company more commonly known by its initials. In 1981 HRH Prince Michael of Kent became a non-executive director. On 23 March 1982, ITT sold on the London market another 10 per cent of its shareholding in STC, reducing its stake to 75 per cent. The placing price was 496p, only 6 per cent below the opening market price that day of 528p, and nearly £50 million was raised within an hour. So enthusiastic were the institutions to whom the shares were offered that their applications had to be scaled down.

That turned out to be merely a prelude to further changes. During what seemed a quiet summer two matters were being deliberated that were fundamental to the start of the company's second century: one of its main product lines, and its ownership. The two separate issues, which to the outside world appeared to be related, were resolved within days of one another in October 1982.

On Monday 4 October it was announced that, after more than ten years of collaboration with its competitors and major customer, STC had withdrawn from the System X programme. The next day its shares fell 17p to 580p in an immediate reaction that the company had been the loser. On consideration, the situation was less serious and perhaps even a blessing in disguise. British Telecom had been under government pressure to rationalise work on System X and produce more competition among its suppliers. Their relationships had not been entirely harmonious, especially on export policy. The product, highly engineered for the well-developed UK network and therefore comparatively expensive, had as yet failed to open up any significant overseas market. There was a small order for the island of St Vincent; but India, for example, had been lost to the French. STC was on the point of signing a loss-leading contract with China for Guangdong province but, overall, export prospects were uncertain. One outside critic described System X as 'Concorde without wings'.

Within STC it was expected to represent 7.7 per cent of turnover and not more than 4 per cent of profit at the end of its current five-year plan. Withdrawal would be immediate. The company was invited to fulfil its contract to deliver the large local exchange by the autumn of 1983 so the rundown would be spread over 12 months. By way of compensation, as GEC and Plessey phased out production of TXE4 exchanges to concentrate on System X, STC would get progressively larger orders for TXE4A. These would amount to at least another £100 million over the next five years, with negligible research and development cost. Some 500 highly-skilled staff engaged on these activities for System X would be redeployed. There are various possibilities. One major shift in the company's business that has already been identified is the growing importance of the range of peripheral equipment being attached to the network, of the end-user in relation to the traditional one major customer. Moves had already been made into word processing and information transfer technology. Within this expanding area there are several lines of development that can be pursued. What the company must do is single out those in which it can succeed and profit.

A fear behind the initial stock market reaction to the news of the System X withdrawal was that by about 1988 STC would find itself out of the public switching business. It is always possible that there could be a demand for the ITT System 12. In that event, the technology could be transferred quickly. Most of ITT's one billion dollar investment in the system is in Europe. Taking a longer term view, STC has the resources to work on the next generation of systems, those to succeed System X and System 12. By 1990 it estimates that its activities will be evenly divided between those that are mainly intellectual and those that are predominantly manual.

Much of this will centre on the creation of increasingly intelligent devices. As well as a range of terminals performing sophisticated functions there will be a dispersal of intelligence through the network. It will not be centred, as at present, in exchanges. With the growth of wideband transmission on optical fibres, the importance of the switching function will diminish. Instead the concepts of distributed processing and message routeing will become important, information being encoded for delivery to specific numbers or addresses.

Few were able to sit back and take such a perspective in the flurry of rumour and activity during that week of October. On the weekend before the System X news broke, speculation suggested that ITT was sounding City of London opinion on offering for sale another 40 per cent of its stake in STC. Market reaction on the Monday was to mark the shares down from 628p to 597p. Immediately followed by the System X news, the situation looked uncertain and the share price fell to 545p.

On Friday 8 October the speculation was shown to have been well-founded when the offer for sale was formally announced of 40 million shares at 525p each. This time shares were reserved for pensioners as well as employees. Underwritten by S. G. Warburg and three other merchant banks, the large offer was more than 13 times over-subscribed, and reduced ITT's stake in STC to 35 per cent. This figure had been agreed between the two companies in July and notified to the British government and British Telecom at least three weeks before the outcome of the System X negotiations.

At the approach of its second century STC was now a majority-owned British company. Its indisputable new status solved a problem of identity that was as old as the company and removed an ambiguity that competitors had always been able to exploit. They had argued that STC was a Trojan horse for ITT interests, that it was not truly British but only a front for the Americans. In the circumstances STC felt that it did not necessarily enjoy the business confidence that its technical and other merits deserved. This was important both in its relationship with British Telecom, which represented 60 per cent of its business, and in dealings with overseas customers. The company was the biggest exporter of telecommunications equipment from the UK.

Nevertheless STC, although no longer an ITT subsidiary, would continue to benefit from a continuing association with ITT, notably in access to worldwide technology. Some of this is in non-telecommunications activities that could become of value. For its part ITT immediately gained $350 million, a contribution to relieving its debt burden of $4.37 billion. It could also look to receiving an increasing return on its reduced investment. Within a few years ITT expected to obtain more from its 35 per cent stake than it did from its 75 per cent. The sale thus served its own long-range purpose.

At its centenary STC employees and shareholders can look at the past and present with satisfaction. To mark the occasion ITT, for the first time in its history, is holding its annual shareholders' meeting outside the USA, in the City of London. At the same time STC is sponsoring a major exhibition in the Science Museum that will become a permanent telecommunications gallery, and in the country at large people will have an opportunity to hear about the technology and its future in the Faraday Lecture being delivered for the Institution of Electrical Engineers. Under Sir Kenneth Corfield (1981 salary £121,752), the second longest serving chief executive since the company was formed, STC has largely made the transition from an electromechanical to an electronic culture. Its new status has given him the kind of further challenge he needs.

In a dynamic electronic culture and a more competitive world, the basis of success is a high rate of innovation which, paradoxically, presupposes a stable environment. Here the Japanese are at an advantage in enjoying a stability of talent, whereas British industry can suffer through the movement of engineers from one organisation to another. That has the benefit of keeping ideas flowing but it seldom results in the consistent application necessary for major technical advances, which usually take eight to ten years and the work of a team rather than a gifted individual. To foster the latter STC must continue to evolve as an open and meritocratic society in which talent is patently recognised and rewarded. The environment it creates has to encourage the kind of pioneers who made the technical contributions that have been the foundation of its success to date. If their like is lost, the company will be at a considerable disadvantage against the enormous teams working away in places the size of Bell Labs or the expanding legion of entrepreneurial innovators enjoying rapid economic growth.

More is likely to happen to the company in the next few decades than in the whole of its first century. Futurologists see a greatly expanded role for telecommunications in a world that will be both energy-hungry and more leisured. As Alec Reeves had forecast, it will be easier and less expensive to move information instead of people. More information will be beamed and cabled into their homes in the form of more TV channels, telephoned instructions to operate cookers or the central heating, and computer-based data on goods and services that will be chosen from the comfort of an armchair. Homes themselves will become environments controlled by microprocessor-based systems. Looking farther ahead, in the twenty-first century there are bio-electronic possibilities, a quantum leap like the direct connection of man and computer. To some it is a frightening prospect.

On the eve of 1984, though, the concern is not so much about a Big Brother state harnessing technology to keep people in submission as about the continuing economic decline of the country. When Western Electric established its agency in 1883 the decline was relative. A century later it is much more obvious and there is discussion on whether the relative has become absolute, whether Britain is a post-industrial society. The contrary argument is that the country is going through another industrial revolution in which the old, heavy and traditional industries are dying and new, young industries like electronics are coming up to replace them. In that sense STC is one of the larger companies creating new technology and is an example of hope for industrial regeneration based on the emerging information society.

Acknowledgements

Of the many people who have contributed to this history I owe most to Miss Laurie Dennett, the STC archivist, who has gathered information principally from the company's written sources, the Post Office, ITT and Western Electric in New York, and Bell Telephone Manufacturing Company in Antwerp.

Interviews with people inside and outside the company, some of which were done by Jack Read, fortunately spanned most of the period. There were so many that individual contributions are not always evident from the text. I have also enjoyed conversations with Dr Robert Sobel, professor of business history at Hofstra University, New York, who coincidentally has been writing the history of ITT. While information has come from a variety of sources its interpretation is my own.

Among UK publications consulted were

City of London Directory 1883–1902
The Electrician
Electrical Review
National Telephone Journal
Stock Exchange Year Book
Stratford Express

Of the writings of J. E. Kingsbury I have mostly drawn on

The Telephone in Principle and Practice (1882)
The Future of the Telephone in the United Kingdom (1902)
The Telephone and Telephone Exchange (1915)
The Story of the Telephone (1916)

The Story of the Telephone by J. H. Robertson (1947) is a useful history of the telecommunications industry in the United Kingdom. Maurice Deloraine's *When Telecom and ITT Were Young* (1976) was supplemented by personal memories.

US publications included

Electrical World
International Communications Review
International Telephone Review

House publications of most value were

STC	*Standard News*
	STC News
ITT	*Electrical Communication*
	International Review
Western Electric	*Western Electric News*

Organisations that helped included

BBC Written Archives Centre
Brighton Museum
The British Library
Cable Makers' Association
Engineering Employers London Association
Institution of Electrical Engineers
Modern Records Centre, University of Warwick
City of Portsmouth Record Office
Royal Military Academy, Sandhurst
Tunbridge Wells Library, Record Office
University of London

In the preparation of the MS I am grateful to Data Recall for the loan of a Diamond word processor and particularly for the assistance of Mrs Valerie Wilkinson and Miss Annette DeVoil.

Accounts: 1883–1981

The following figures should be regarded as only an indication of financial trends. They cannot be compared year on year because of changing legal entities, accounting practices and taxation policies.

Figures in English Pounds. Rate 1919 – $4.85=£1.

WESTERN ELECTRIC COMPANY, LONDON

Year(s)	*Net sales*	*Net profits on investment*	*Average investment*	*Percentage return on (1) Sales*	*Percentage return on (2) Investment*	*Number of employees*	*Notes*
1883–1890	264,340	24,953	17,000	9·4	18·0	—	
1891–1895	323,632	33,252	27,000	10·1	18·1	—	
1896–1900	795,232	33,021*	82,000	2·1	8·1	473	*Special plant depreciation of $100,000 written off in *both 1900 and 1901*. 1901 includes 14-month figures for Johannesburg.
1901–1905	1,986,802	18,508*	311,000	1·3	1·2	895	
1906	550,408	16,580	484,000	3·0	3·4	941	
1907	724,701	54,263	537,000	7·5	10·1	837	
1908	619,508	56,599	519,000	9·2	10·9	1,717	
1909	479,413	26,074	466,000	5·4	5·6	975	Includes 13-month figures for Johannesburg.
1910	492,181	25,621	289,000	5·1	6·6	1,617	Ditto. First year WECo, London a limited company
1911	413,863	84,327	461,000	19·5	18·3	3,098	

WESTERN ELECTRIC COMPANY, LONDON – *continued*

Year(s)	*Net sales*	*Net profits on investment*	*Average investment*	*Percentage return on (1) Sales*	*(2) Investment*	*Number of employees*	*Notes*
1912	526,825	22,759	604,000	4·3	3·8	2,569	
1913	715,336	12,065	672,000	1·5	1·8	2,331	
1914	701,134	10,316	699,000	1·3	1·5	2,388	
1915	774,351	66,750	674,000	9·5	9·9	1,703	
1916	897,171	102,020	618,000	11·2	16·5	1,941	
1917	1,126,986	101,075	539,000	8·7	18·8	1,894	
1918	1,022,494	88,526	536,000	8·4	16·5	1,885	
1919	910,142	85,876	689,000	9·2	12·5	3,156	
1920	1,673,900	102,911	1,009,000	6·2	10·2	1,420	
1921	3,271,000					3,121	Includes Tokyo.
1922	2,110,000					2,810	
1923	2,233,000					3,331	
1924	3,014,000					5,206	
1925	3,514,000					6,318	

Notes: (1) Figures include those for Sydney until 1909. Contract service charge, general expense and general interest are excluded.
(2) Rates prior to 1907 are based on total sales to customers and other houses; for 1907 and 1908, on sales to customers plus sales of WE merchandise to other houses; for subsequent periods, on sales to customers only.

STANDARD TELEPHONES AND CABLES LTD

Year	*Sales (£000s)*	*Net income before tax (£000s)*	*Taxation (£000s)*	*Net income after tax (£000s)*	*Dividend (£000s)*	*Employees*	*Notes*
1926	3,466		75	460	360	6,267	
1927	3,287			365		6,053	
1928	3,066			406		5,578	
1929	3,263			422		6,431	
1930	3,990			576		7,718	
1931	2,554			36		5,449	
1932	2,366			(50)		5,663	
1933	2,240			79		4,912	
1934	2,447	242	34	208		6,507	
1935	3,311	557	90	467	388	7,911	
1936	4,199	795	254	541	297	8,837	
1937	5,039	887	255	632	293	10,779	
1938	6,490	1,383	457	926	493	11,377	
1939	6,492	942	523	419	487	12,199	
1940	7,704	657	409	248	695*	14,448	*Including capital issue.
1941	9,506	1,008	649	359	—	18,953	
1942	12,955	1,184	696	488	125	23,718	
1943	14,346	1,193	749	444	188	25,287	
1944	14,024	1,070	664	406	127	23,794	
1945	12,256	525	354	171	187	16,379	
1946	10,096	858	452	405	206	15,447	
1947	10,449	616	265	351	208	16,906	
1948	15,285	1,188	624	504	512	17,500	

STANDARD TELEPHONES AND CABLES LTD – continued

Year	*Sales (£000s)*	*Net income before tax (£000s)*	*Taxation (£000s)*	*Net income after tax (£000s)*	*Dividend (£000s)*	*Employees*	*Notes*
1949	16,369	1,640	968	672	275	17,985	Dividend restraint agreed with Federation of British Industries.
1950	16,964	1,920	1,100	820	269	18,743	Dividend restraint as for 1949.
1951	19,538	2,181	1,201	980	263*	20,758	*Capital increased to £5·5 m.
1952	25,204	2,953	1,893	1,060	361	22,606	
1953	27,807	4,021	2,640	1,381	445	22,351	
1954	25,894	3,165	1,585	1,580	578*	21,360	*Capital increased to £7 m.
1955	29,141	3,363	1,534	1,829	535	22,900	
1956	32,581	3,015	1,317	1,698	798*	24,303	*Capital increased to £9 m.
1957	34,128	2,172	827	1,345	798	24,693	
1958	30,447	553	52	501	867*	24,402	*Capital increased to £11·2 m.
1959	30,083	689	49	640	956*	23,726	*Capital increased to £12 m.
1960	33,760	498	(68)*	430	—	24,155	*Tax refund £68 m.
1961	39,333	1,442	(10)	1,452	—	25,280	
1962	45,581	1,911	581	1,330	943	27,160	
1963	61,256	4,148	1,478	2,670	1,415	28,811	
1964	66,746	5,793	2,309	5,047	2,265	29,935	Extraordinary credit 1,563.
1965	64,405	5,241	1,150	4,091	2,045	c. 31,000	
1966	78,817	7,111	2,568	4,543	2,157	32,750	
1967	86,947	9,397	3,815	5,582	2,465	33,194	
1968	106,954	7,439	3,537	3,902	1,522	34,140	
1969	122,612	6,446	3,322	3,124	1,000	33,053	
1970	140,300	5,655	1,987	3,668	1,000	35,393	
1971	161,447	11,828	4,355	5,740	2,000	33,608	Extraordinary items 1,733.
1972	187,668	21,894	8,112	13,618	4,000	33,264	Extraordinary items 164.

STANDARD TELEPHONES AND CABLES LTD – continued

Year	*Sales (£000s)*	*Net income before tax (£000s)*	*Taxation (£000s)*	*Net income after tax (£000s)*	*Dividend (£000s)*	*Employees*	*Notes*
1973	286,800	38,956	15,965	22,991	8,729	39,067	
1974	332,971	33,900	17,832	16,068	8,000	39,370	
1975	382,900	25,855	14,117	10,452	8,500	41,338	Extraordinary items 1,286.
1976	431,839	28,088	13,166	9,440	9,300	40,500	Extraordinary items 5,482.
1977	460,083	31,099	4,947	27,345	9,000	37,000	Extraordinary credit 1,193.
1978	508,719	29,765	3,847	24,795	8,000	36,000	Extraordinary items 1,123.
1979	436,862	33,433	5,895	27,538	8,000	27,956	
1980	537,700	44,092	15,442	28,650	10,000	27,300	
1981	567,500	50,600	14,100	36,500	13,500	25,000	

Index